CONTENTS

ENVIRONMENT AND HERITAGE SERVICE

JOINT NATURE CONSERVATION COMMITTEE

DEPARTMENT of ARTS CULTURE and the GAELTACHT

Institute of Terrestrial Ecology

Atlas of grasshoppers, crickets
s in Britain

Centre for Ecology and Hydrology

London: The Stationery Office

Natural Environment Research Council

The Institute of Terrestrial Ecology (ITE) is a component research organisation within the Natural Environment Research Council. The Institute is part of the Centre for Ecology and Hydrology, and was established in 1973 by the merger of the research stations of the Nature Conservancy with the Institute of Tree Biology. It has been at the forefront of ecological research ever since. The six research stations of the Institute provide a ready access to sites and to environmental and ecological problems in any part of Britain. In addition to the broad environmental knowledge and experience expected of the modern ecologist, each station has a range of special expertise and facilities. Thus, the Institute is able to provide unparallelled opportunities for long-term, multidisciplinary studies of complex environmental and ecological problems.

ITE undertakes specialist ecological research on subjects ranging from micro-organisms to trees and mammals, from coastal habitats to uplands, from derelict land to air pollution. Understanding the ecology of different species of natural and man-made communities plays an increasingly important role in areas such as monitoring ecological aspects of agriculture, improving productivity in forestry, controlling pests, managing and conserving wildlife, assessing the causes and effects of pollution, and rehabilitating disturbed sites.

The Institute's research is financed by the UK Government through the science budget, and by private and public sector customers who commission or sponsor specific research programmes. ITE's expertise is also widely used by international organisations in overseas collaborative projects.

The results of ITE research are available to those responsible for the protection, management and wise use of our natural resources, being published in a wide range of scientific journals, and in an ITE series of publications. The Annual Report contains more general information.

> The Biological Records Centre is operated by ITE, as part of the Environmental Information Centre, and receives financial support from the Joint Nature Conservation Committee. It seeks to help naturalists and research biologists to co-ordinate their efforts in studying the occurrence of plants and animals in the British Isles, and to make the results of these studies available to others.

P T Harding
Institute of Terrestrial Ecology
Monks Wood
Biological Records Centre
Abbots Ripton
Huntingdon
Cambs PE17 2LS

Tel: 01487 773381   Fax: 01487 773467
WWW: http://www.nmw.ac.uk/ite

Introduction

Recording the distribution of orthopteroids[1] in Britain and Ireland

The early history of recording in Britain and Ireland is described by both Ragge (1965) and Marshall and Haes (1988). The earliest attempts to collate information on the distribution of species were probably those of Shaw (1889, 1890) and Burr (1897), who listed some localities, and Lucas (1920) with more detailed records. These were followed, at a more popular level, by Burr (1936) and Pickard (1954). Kevan published a series of papers on the distribution of orthopteroids in Britain and Ireland (see especially Kevan 1952 and 1961). Burr (1936) included county maps for most species in Britain and Ireland and Ragge (1965) published maps of the vice-county distribution of most species, which were updated in 1973 (Ragge 1973). Marshall (1974) drew extensively on this resource of information for her review of changes in the British orthopteroids since 1800.

Recording in England

As indicated in the historical survey by Marshall and Haes (1988), field studies of orthopteroids in England include observations by Thomas Moffet in the late 16th century and the familiar observations on crickets by Gilbert White in the late 18th century. Specimens of several species collected in the first quarter of the 19th century, from the collections of J F Stephens and his contemporaries, are in several museums, particularly the Hope Entomological Collections at the University Natural History Museum, Oxford. During the 19th century two celebrated entomologists, J C Dale (father), in the first half, and C W Dale (son), in the second half, included orthopteroids in the wide range of the insects they studied and collected.

Several other orthopterists were recording and collecting by the end of the last century and first decade of the present one. The efforts of over 30 recorders in England were summarised by W J Lucas (1920). Lucas himself was an active field worker and provided an invaluable tally of records (both his own and those of others), mainly in *The Entomologist*, until 1930. Possibly because grasshoppers and related insects do not make particularly attractive collections, few other entomologists worked on them during the 1920s and 1930s. Exceptions included the comprehensive survey of the South Haven peninsula in Dorset by C and P Diver, and the work of M D Burr after retirement.

In the 1940s a brilliant young scientist, E J Clark, took up the study of grasshoppers as a recreational activity. His tragically early death (in 1945) probably denied us of a worthy successor to Lucas, at a period just before the major changes in the countryside were brought about through technological developments in agriculture. From the mid-1940s another young enthusiast, Ian Menzies, made such important discoveries as the first mainland records of *Conocephalus discolor* (in Sussex in 1945 and Dorset in 1947), and *Metrioptera roeselii* in Surrey in 1947. Ian Menzies is still an active recorder, certainly the longest-serving orthopterist in England this century. However, David Ragge has been active for almost as long and, with Bernard Pickard, he provided many important records during the 1950s, by which time D K McE Kevan, R M Payne, John Burton and others were covering extensive areas of the country for orthopteroids. Ragge's book in 1965, followed soon after by the launch of the Orthoptera Recording Scheme, ensured that, from the end of the 1960s, there was a substantial body of recorders of orthopteroids in England. Colour photography was an important objective for some recorders, which was aided by the general availability of fast colour film and developments in camera equipment.

Special recognition for an outstanding contribution to the field recording of orthopteroids in England, since the start of the scheme, must go to Michael Skelton. Without in any way intending to devalue the efforts of other recorders (see **Acknowledgements**), particular mention should also be made of the contributions by Keith Alexander (of the National Trust's Biological Survey Team), John Burton, John Paul and, more recently, Chris Timmins. In addition to ongoing recording of orthopteroids by several local biological records centres, detailed surveys aimed at the production of comprehensive, new, county orthopteroid atlases (some replacing existing atlases) are currently being organised by David Baldock (Surrey), Robert Cropper

---

[1] *Throughout the* **Introduction** *the term 'orthopteroid(s)' is used to include all the groups covered in this* Atlas

1

(Somerset), Chris Haes (Cornwall), David Haigh and Alan Wake (Gloucestershire), Jennifer Newton (Lancashire), David Richmond (Norfolk), Richard Surry (Dorset), David Veevers (Kent), Alan Wake and others (Essex), and John Widgery (Hertfordshire).

Recording in Scotland

W J Lucas (1920) included many Scottish records, some going back to the previous century, including those from the enthusiastic A M Stewart and key data from a paper by W Evans (1901). Amongst the earliest references to a British orthopteran was that by Sir Richard Sibbald (in 1684) to *Gryllotalpa gryllotalpa*, although it is not clear where in Scotland it occurred. Some early surviving specimens of Orthoptera are in Perth Museum, collected by F B White in the north-west Highlands and Kirkcudbrightshire in 1868 and 1870. Burr (1936) was not very informative about Scotland, but Kevan (1961) provided data which enabled Ragge (1965) to produce vice-county maps which are still applicable. From the 1970s many visiting recorders have added data, particularly for the Highlands and the Hebrides. Key island records by resident recorders have been provided by Jane Dawson from Islay, Tristan ap Rheinallt from Arran, and Keith Fairclough from Orkney. Over the last decade Stephen Moran at Inverness Museum, Ian Francis from the Royal Society for the Protection of Birds Grampian region and Stephen Hewitt formerly at Perth Museum have provided key data for the northern mainland, and Jim McCleary for the south-west of Scotland.

Recording in Wales

Until the 1980s, surprisingly little had been published on Welsh orthopteroids. Even Lucas (1920) and Burr (1936) were relatively uninformative. Data collected by Kevan (1961) and used to provide county maps by Ragge (1965) indicated that the country was distinctly under-recorded. However, some important new records were made by John Burton in the late 1960s (Burton 1971). An event that encouraged orthopterists to take an interest was the remarkable discovery of *Metrioptera roeselii* in 1970 (Ragge 1973). Fortunately, in the course of other studies, Joan Morgan has provided a flow of Welsh orthopteroid records, since the inception of the national orthopteroid recording scheme. Some records from the insect-rich Gower were given in Mary Gillham's (1977) natural history of the peninsula, to be followed by detailed records from Ian Tew in the 1980s. Records from

Carmarthenshire were compiled by Ian Morgan in the same period, Alan Wake recorded in Montgomeryshire during 1988–90, and John Steer has provided annual lists for Pembrokeshire since 1988. Another important contribution was the unpublished list of orthopteroids for 'west Wales' provided by John Comont in 1985, on behalf of the Dyfed Wildlife Trust. The first comprehensive paper on Welsh orthopteroids, with 10 km square maps for the entire south-west of the country, was by Adrian Fowles in the *Dyfed Invertebrate Newsletter* (1986). This has been updated in subsequent newsletters published by the Dyfed Invertebrate Group.

Recording in Ireland

The distribution of orthopteroids in Ireland did not attract much attention during the late 19th and early 20th centuries, when so much valuable work was done on other insect groups by entomologists such as G H Carpenter, J N Halbert, E O'Mahony and A W Stelfox. Following the publication of Ragge (1965), Forsyth (1968) published a request for records, but nothing seems to have come of this initiative. Interest in the group began to show again in the 1970s, and Speight (1976) published a number of new records, including the first for *Tachycines asynamorus*. Don Cotton (1980, 1982) collated his own records with those of other entomologists in Ireland (including Martin Speight, Des Higgins and Jim O'Connor) and museum material. Ryan, O'Connor and Beirne (1984) listed publications relating to orthopteroids in Ireland published up to 1980. Other entomologists, including some visitors from Britain, have added scattered records since the early 1980s. During the past 25 years, four species have been added to the Irish list (*Tachycines asynamorus* in 1975, *Metrioptera roeselii* in 1977, *Pholidoptera griseoaptera* in 1983 and *Conocephalus dorsalis* in 1989). *Chorthippus albomarginatus*, which was first recorded in 1960, has been confirmed as an Irish species and subsequently has been found at a number of sites. Several introduced species have been recorded for the first time in Ireland over this period.

Recording in Isle of Man

Although little was published on Manx orthopteroids in the pre-1961 period, several important finds had been made by local entomologists, particularly the discovery of *Leptophyes punctatissima* at Perwick in 1924, where it still occurs. The island came to the attention of orthopterists with the discovery of *Stenobothrus stigmaticus* at its only British location, on the Langness peninsula, by R W Crosskey in 1962 (Ragge 1963). Since the inception of the national recording scheme, Larch Garrad, formerly of the Manx Museum, has provided the scheme with an almost annual summary of the orthopteroids of the island. This has been supplemented by occasional records and published papers by visiting recorders, particularly in connection with the *S. stigmaticus* (eg Burton 1990; Cherrill 1994).

Recording in Channel Islands

The earliest detailed summaries of orthopteroids were papers by W A Luff (1896) for Guernsey and, for the entire archipelago, by Burr (1899). These insects were subsequently reassessed, after a visit in 1938, by F E Zeuner (1940). Bernard Pickard (1954) wrote a special Channel Island section for his book on British Orthoptera, which included key records made in the late 1940s by D J Clennet and W J le Quesne, both of whom continued to add records during the next 20 or more years. Some of their later records were included by Frances Le Sueur (1976) in her book on Jersey. Mark Amphlett provided data from Guernsey for 1979–83, updating Luff's records, while John Paul updated Zeuner's paper with his own (1994), based on visits to Jersey in 1981 and 1991. From the early 1970s, Roger and Margaret Long have continuously maintained orthopteroid records for the islands.

The Biological Records Centre and recording

The collection, by individuals, of detailed records of orthopteroids had been under way for decades, but their work did not receive a national focus until the late 1960s as a result of the setting up of the Biological Records Centre (BRC) in 1964 (Harding & Sheail 1992). Recording of invertebrates began in earnest at BRC in 1967 with the appointment of the late John Heath as its zoologist. Recording of Orthoptera and allied orders formed part of the Insect Distribution Maps Scheme launched in 1967–68, and in its early years recording of orthopteroids was run in tandem with the recording of Odonata (see Merritt, Moore & Eversham 1996).

Working as assistant to John Heath at BRC, Michael Skelton took special responsibility at BRC for orthopteroids and Odonata for a number of years and edited the data for the first *Provisional atlas* of 10 km square maps of orthopteroids published in 1978 (Skelton 1978). The maps in this *Atlas* were something of a revelation because for most species they accurately (if somewhat

patchily) demonstrate the range of most species. Inevitably, they are much more meaningful than the coarse scale of the earlier vice-county maps in Ragge (1965).

In 1977 Chris Haes assumed responsibility for running the Orthoptera (and allied orders) Recording Scheme, on behalf of BRC, as an independent, voluntary scheme organiser. He collaborated with Paul Harding at BRC to prepare a second edition of the *Provisional atlas* at the end of the 1979 field season (Haes 1979). The scheme has continued under Chris Haes' leadership until he handed over to John Widgery in October 1995 (see below). Up-to-date information on orthopteroids in Britain (and Ireland) has been published regularly by Chris Haes in the bi-monthly journal *British Wildlife* since June 1990.

Up until 1980, almost all the computerised records held at BRC (including those for orthopteroids) were only 10 km summaries. As part of a general move to improve the usefulness of data held by BRC, Paul Harding, the late Dorothy Greene and Chris Preston initiated a programme to computerise complete records wherever possible (see Harding & Sheail 1992). This programme has continued, with Brian Eversham taking joint responsibility with Paul Harding for overseeing work on invertebrates (including Orthoptera) soon after he joined BRC in 1983. More recently, Julian Dring, and subsequently Henry Arnold, have been responsible for managing the computerised data for orthopteroid orders at BRC, most of which were entered on to the computer by Wendy Forrest and Val Burton.

Identification guides

Formalised recording of orthopteroids for the BRC scheme was greatly aided in the early years by the existence of an up-to-date, authoritative and well-illustrated account of British and Irish Orthoptera and allied orders (Ragge 1965) and a companion gramophone record of their songs, produced in association with the British Broadcasting Corporation. The book contains vice-county distribution maps for native species and brief accounts of the occurrence of many accidentally introduced non-native species.

A volume was published in the *Naturalists' Handbooks* series (Brown 1983), which includes keys to the native grasshoppers, crickets and bush-crickets (but not ground-hoppers or the

Table 1. Local distribution atlases of Orthoptera (and allied orders) published since 1975

County (vice-county number)		Publication
ENGLAND	Bedfordshire (30)	Rands (1978 onwards)
	Berkshire (22)	Paul (1989)
	Buckinghamshire (24)	Paul (1989)
	Cornwall (1&2)	Haes (1990)
	Derbyshire (57)	Frost (1991)
	Devon (3&4)	Davies (1987)
	Dorset (9)	Mahon (1992)
	Essex (18&19)	Wake (1984)
	Hampshire (New Forest area) (11)	Welstead & Welstead (1985)
	Hertfordshire (20)	Widgery (1991)
	Norfolk (27&28)	Richmond & Irwin (1991), Richmond (1995)
	Oxfordshire (23)	Paul (1989)
	Somerset (5&6)	Cropper (1993)
	Sussex (13&14)	Haes (1976)
	Yorkshire (Sheffield area) (63)	Whiteley (1981)
	Warwickshire (38)	Copson (1984)
WALES	Ceridigion (46)	Fowles (1986, 1992)
	Dyfed (44&45)	Fowles (1986)

other orthopteroid orders) with coloured plates of all but the rarest species. It was revised in 1990 and is still in print in 1996.

The maps produced by Chris Haes for the second edition of the *Provisional atlas* were updated in 1988 by Harley Books, for inclusion in the new handbook *Grasshoppers and allied insects of Great Britain and Ireland* (Marshall & Haes 1988). This book drew heavily on information accumulated by Chris Haes as a result of the recording scheme. It is still in print and is the essential companion to the present *Atlas*. In an attempt to avoid duplication, each species account in this *Atlas* is cross-referenced to Marshall and Haes (1988), where much more detail will be found about the appearance, life history and habitats of species, together with a key to adults and excellent species portraits by Denys Ovenden. Inevitably, the 10 km square distribution maps in Marshall and Haes are now out of date.

In the same year (1988), a Collins *Field Guide* to the Orthoptera (only) of Britain and northern Europe was published (Bellmann 1988). This *Field Guide* covers most British (and Irish) species, with brief species accounts and photographic portraits. This too is still in print and provides a useful supplement to Marshall and Haes (1988), as does the volume on northern European orthopteroids by Holst (1986).

Local atlases and reviews

Marshall and Haes (1988) listed a number of local atlases that were published mainly in the late 1970 and 1980s. Since then more local atlases have been published, although some have been rather provisional in nature and subsequently have been updated, mainly in local journals. Several other updates were known to be in preparation in 1995. Local atlases published since 1975 are listed in Table 1, in alphabetical order by counties, with biological (Watsonian) vice-county numbers in brackets. Although some privately published atlases are listed, those that are listed have been made widely available by the publishers. A few, more ephemeral, 'publications' are not included because they have had very restricted distribution and therefore are not regarded as being in the public domain.

Protection/threat status

Brief reference is made to the protection and threat status of species, for example whether they are listed in legislation, included in a *Red Data Book*, or listed with some other form of status. Several species are protected under legislation in Great Britain and in the Isle of Man. The only published national *Red Data Book* is for Great Britain (Shirt 1987) – local (county) *Red Data Books* are not covered here. Information on the Nationally Scarce status of species (see Ball 1986, 1994) has been provided by the Joint Nature Conservation Committee (JNCC). The additional records summarised in this *Atlas*, together with the adoption by the JNCC of new IUCN (International Union for the Conservation of Nature and Natural Resources) criteria for future British *Red Data Books*, mean that the existing published species statuses quoted here will require re-assessment in future. However, the revision of *Red Data Book* and Nationally Scarce statuses is beyond the scope of this *Atlas* and would not be appropriate here. Three species have been the subject of the Species Recovery Programme operated by English Nature of which one, the mole cricket, is also listed for further action following the publication of the UK Biodiversity Action Plan (Department of Environment 1994) in 1994.

European distribution mapping

Although the orthopteroid faunas of Britain and Ireland are comparatively poor, they are of interest in a broader European context. Species mapping projects which collate data from contributors across parts or all of Europe have been attempted for flowering plants, amphibians, reptiles, birds, mammals and some Lepidoptera. A project to collate data on European orthopteroids has been proposed recently (Erik van Nieukerken pers. comm.).

Atlases, similar to this for Britain and Ireland, have been, or are being, prepared for several parts of western Europe:

Belgium – Devriese (1988)

France – in preparation (Voisin 1992)

Hesse (Germany) – Ingrisch (1979)

Netherlands – Kleukers *et al.* (1997)

Switzerland – in preparation (Thorens & Nadig in prep.)

Holst (1986) summarises the distribution of Orthoptera in northern Europe on a regional basis.

By way of contributing to the gradual process of assembling international summaries, from national and regional sources, small-scale summary maps, showing the distribution of species in 50 km squares of the Universal Transverse Mercator (UTM) grid, are included in this *Atlas*. The UTM grid is the most widely used grid system for mapping at a European scale. These summary maps also have the advantage of 'smoothing' the data from relatively underworked areas, such as parts of Ireland, so that the broad geographic ranges of species may be more readily understood.

Future recording

A national recording scheme for orthopteroids is continuing to be organised in collaboration with the Biological Records Centre. Further details are available on request from the BRC at the address on the inside front cover.

Coverage maps

Coverage maps, summarising all the available records of native and naturalised species, have been included for each of the four groups included in this *Atlas*. Thus, records of vagrant or introduced locusts, non-native cockroaches and exotic earwigs have been omitted completely from the coverage maps. A single map of the coverage of records of naturalised stick-insects is included, but the species have not been mapped individually.

Two overall coverage maps are included, which summarise, for all native and naturalised species of Orthoptera, and native Dictyoptera and Dermaptera, the number of records per 10 km square, and the total number of species recorded in each 10 km square. Records of naturalised stick-insects have been excluded from these overall coverage maps.

Coverage of all species of native and naturalised Orthoptera, and native Dictyoptera and Dermaptera

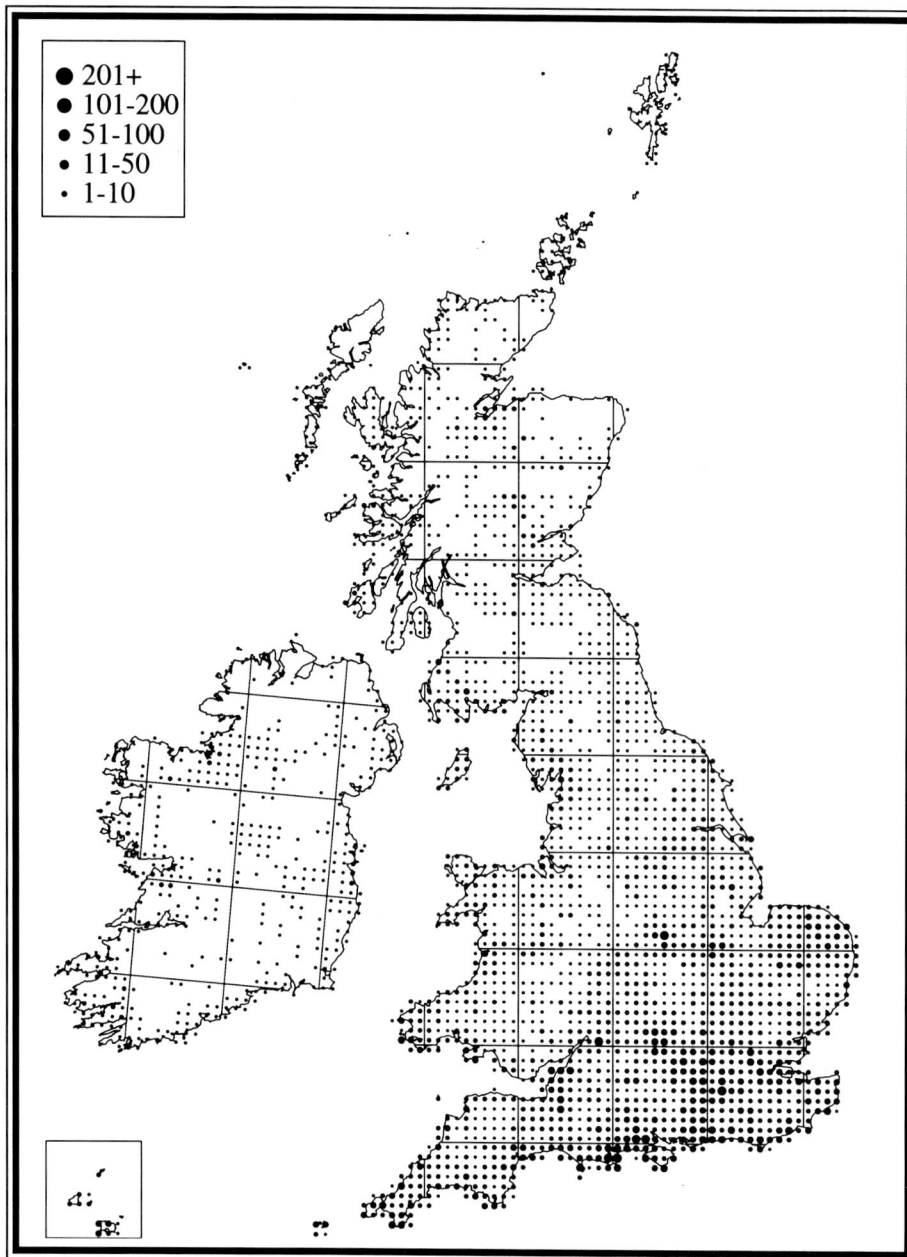

This map shows the coverage in terms of the number of records per 10 km square.

Numbers of species of native and naturalised Orthoptera, and native Dictyoptera and Dermaptera recorded

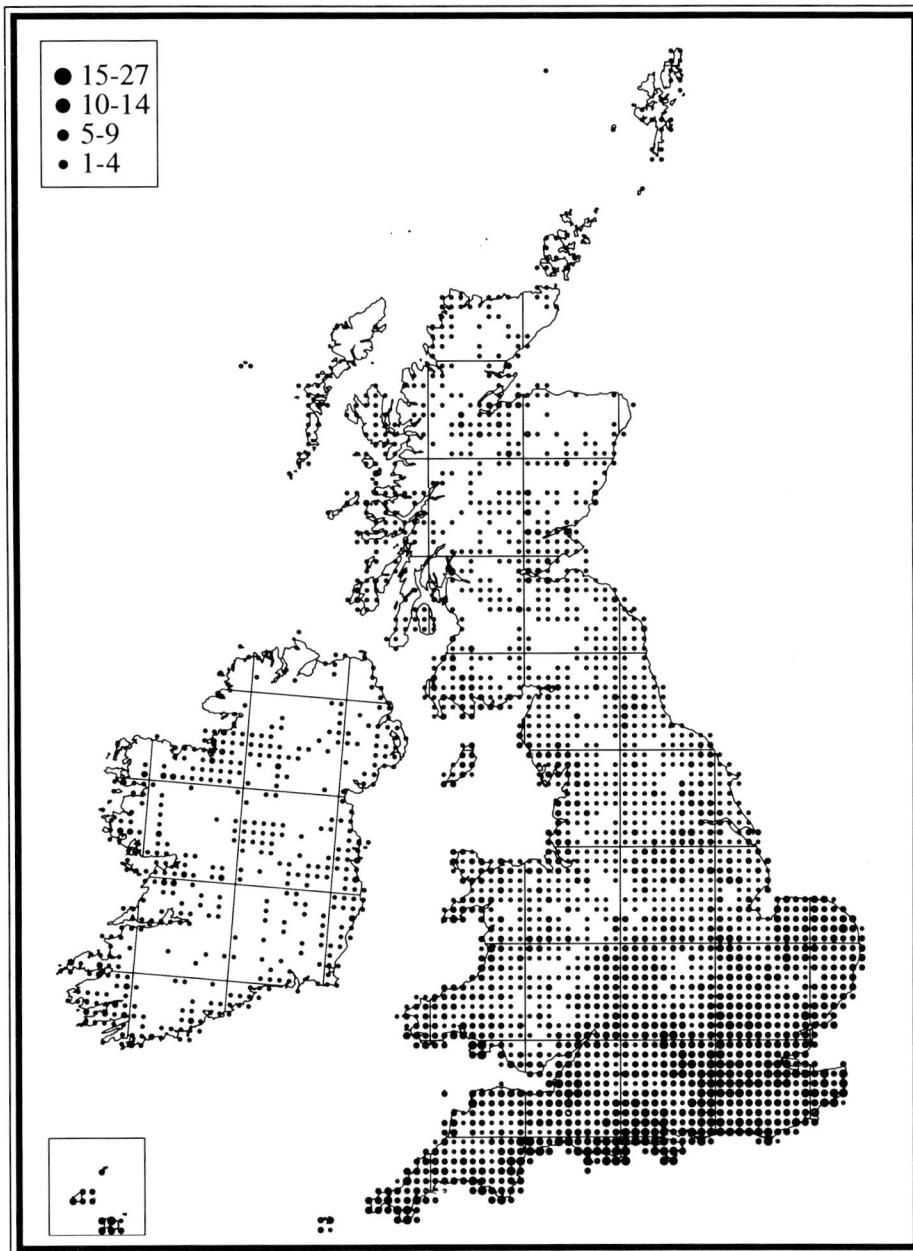

This map shows the total number of species
recorded in each 10 km square.

Orthoptera – grasshoppers and crickets

This is a large and diverse order of insects with a world fauna of over 17 000 described species, most of which occur in the tropical regions, in all types of terrestrial habitats – from desert to rainforest. Only about 900 species occur in western Europe, the great majority in the Iberian peninsula, Italy and the adjacent islands.

The species of Orthoptera that occur in Britain and Ireland are divided into two sub-orders, Ensifera ('crickets') and Caelifera ('grasshoppers'). Within each sub-order, classification at the family level provides six readily recognisable divisions. These systematic divisions are reflected in the common names that have been in general use for a few species for centuries and for others since at least the 1930s:

Ensifera

Raphidophoridae	Camel-crickets[1]
Tettigoniidae	Bush-crickets and cone-heads
Gryllidae	Crickets
Gryllotalpidae	Mole-crickets

Caelifera

Tetrigidae	Ground-hoppers
Acrididae	Grasshoppers and locusts[2]

There are 27 apparently native species recorded as breeding in mainland Britain and 11 species in Ireland, although two additional species have been recorded only in the Channel Islands and one species only in the Isle of Man. However, the true status of some of these species is uncertain, but because they are well established as self-sustaining populations, breeding in natural conditions, they should be regarded as part of our native or naturalised fauna. In addition, there are two alien species that breed only in artificial conditions (see below).

All the native species of Orthoptera are covered here, plus the house cricket (*Acheta domesticus*), which has been found breeding in a few outdoor locations, but occurs mainly in artificial conditions. The greenhouse camel-cricket (*Tachycines asynamorus*) has never been found breeding out of doors or away from artificially heated situations and is not mapped here.

Marshall and Haes (1988) list several additional species that have been recorded as casual or deliberate introductions, or which have arrived naturally as long-distance migrants, probably transported well beyond their normal range by freak weather conditions. These additional species include six species of cricket and four species of grasshopper and locust which are described briefly (pp145–147). Of these, three are illustrated (Plate 5): the desert locust (*Schistocerca gregaria*) (of which over 70 specimens were seen in southern England in October and November 1988), the migratory locust (*Locusta migratoria*), and the Egyptian grasshopper (*Anacridium aegyptium*).

The coverage map opposite includes records of only the native and naturalised species (including *Acheta domesticus*).

---

[1] *Camel-crickets are not native to Britain or Ireland and occur as a result of accidental introductions*
[2] *Locusts are not native to Britain or Ireland, but occasionally arrive as migrants*

Distribution of records of native and naturalised grasshoppers, bush-crickets and crickets

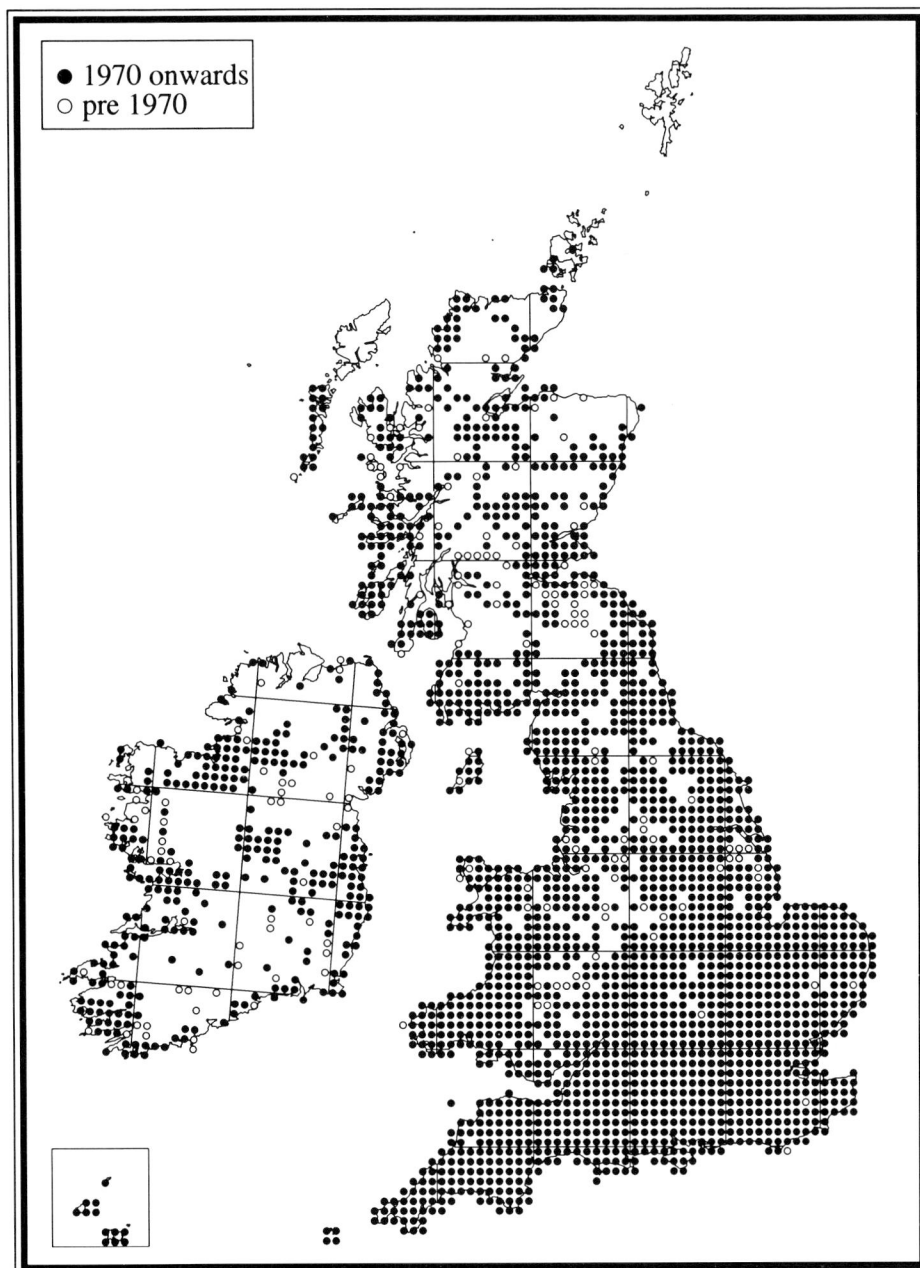

Species accounts and distribution maps

Meconema thalassinum (De Geer, 1773)

Oak bush-cricket

Marshall and Haes (1988) Description and species account pp80–81, Plate 1.

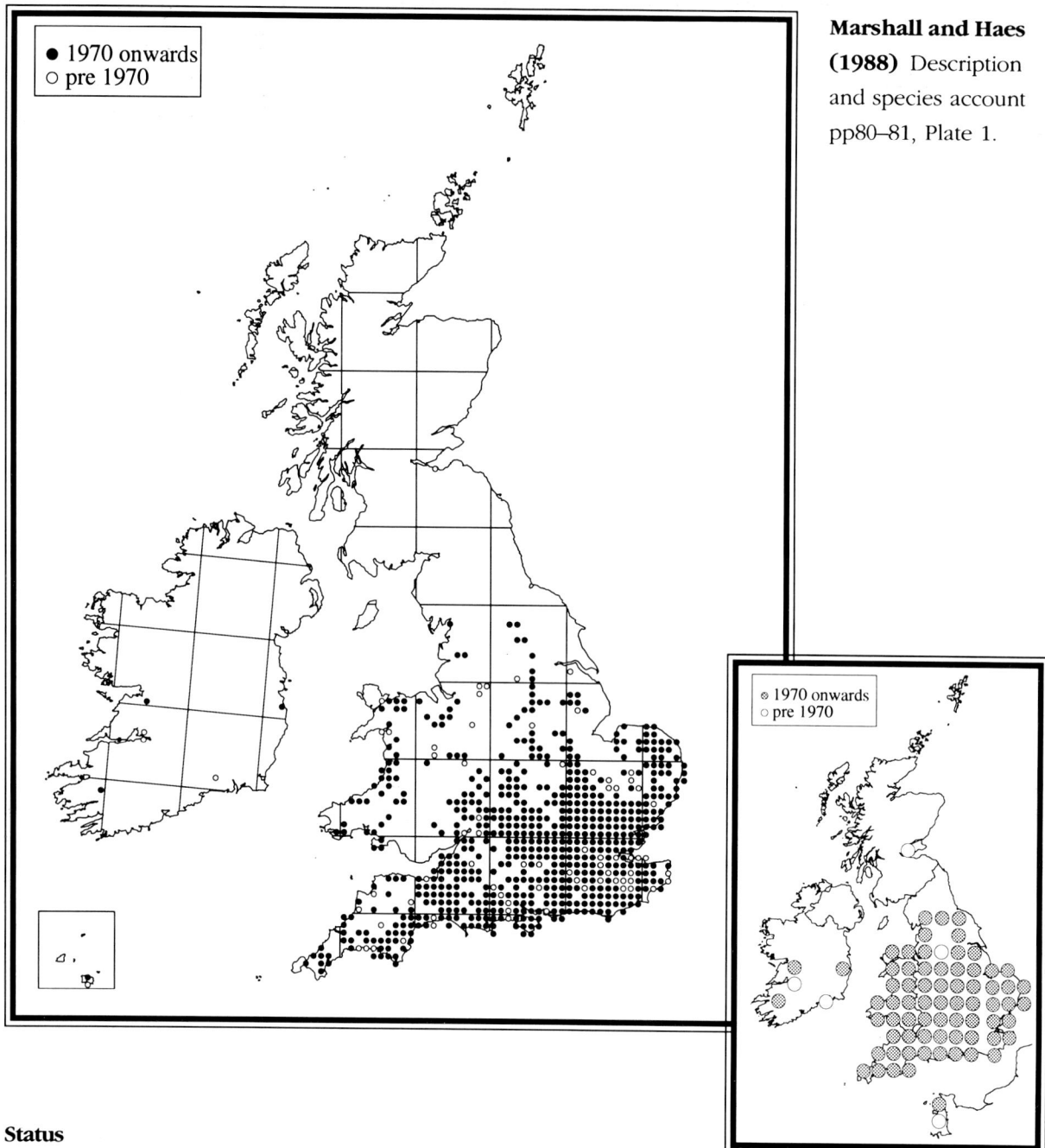

● 1970 onwards
○ pre 1970

⊘ 1970 onwards
○ pre 1970

Status

Native and probably widespread in mature broadleaved woodland in southern Britain and in Jersey. Recorded from scattered woodland localities in southern Ireland – probably native but under-recorded.

This is an elegant pale-green insect with a yellowish dorsal stripe and is one of the smaller bush-crickets (13–17 mm). Both sexes are fully winged and it is the only British bush-cricket which is largely carnivorous. This is probably one of the commonest bush-crickets but, because it does not stridulate (only a resonant tapping with one hind leg on a leaf) and is nocturnal and well camouflaged, it is easily overlooked. Adults are present from late July until late autumn and fly to light at night. It is completely arboreal, occurring mainly in woods, hedges and gardens, and may be beaten from the foliage of shrubs and broadleaved trees in the day. Even early instar nymphs are readily recognisable when collected in this way in late spring and early summer. At the northern edge of its range, in Yorkshire, it is almost exclusively known from sites on magnesian limestone, but otherwise it appears not to be limited to any particular soil or rock types.

Tettigonia viridissima (Linnaeus, 1758) Great green bush-cricket

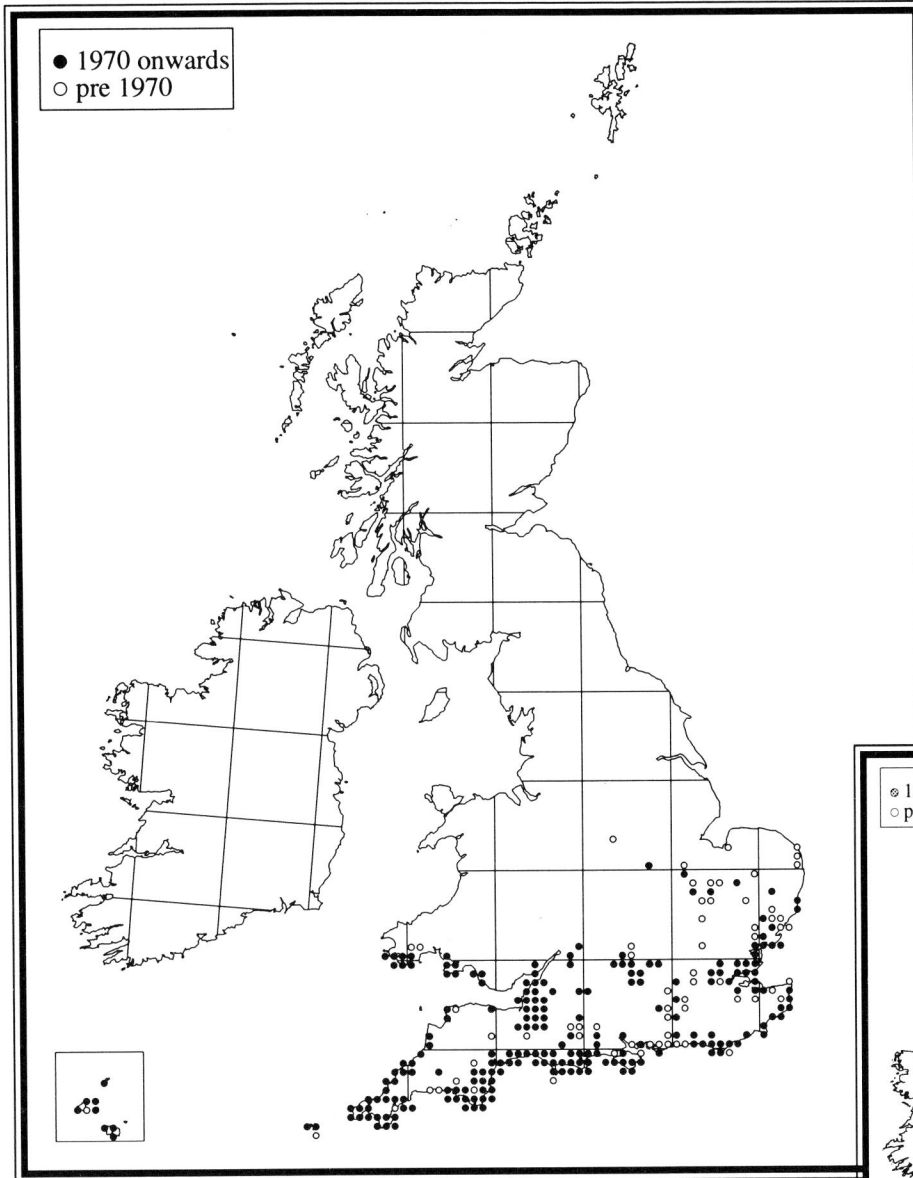

Marshall and Haes (1988) Description and species account pp81–82, Plate 1.

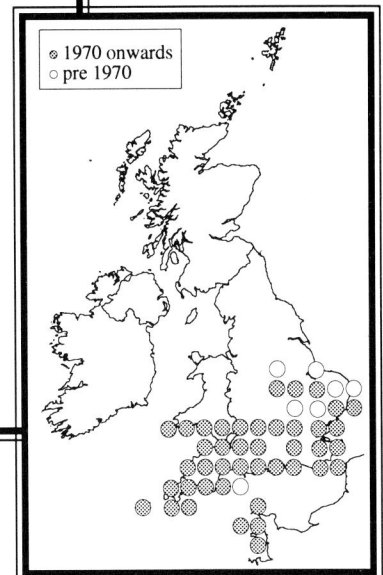

● 1970 onwards
○ pre 1970

⊘ 1970 onwards
○ pre 1970

Status

Native in the southern half of England, particularly in coastal areas, and along the south coast of Wales. Common on the larger Channel Islands. Accidentally imported once into Ireland (Good & Cullinane 1990).

This is our largest orthopteran and one of the largest insects in northern Europe. Despite its large size (40–54 mm), the overall leaf-green colour with a brown dorsal stripe, particularly noticeable in the male, is excellent camouflage in the coarse vegetation and shrubs with which it is usually associated. Stridulation is loud and penetrating – like a shrill computer printer – and can be heard over 100 m or more. It occurs in such places as overgrown hedges, bramble (*Rubus* spp) patches and bracken (*Pteridium aquilinum*) and in

scrub on cliffs; it even occurs in shrubby gardens in some south coast sites. However, thin turf or bare ground are essential for oviposition, and most records are from areas with light or chalky soils. It seems to have become less widespread in inland areas where most surviving populations have been known for several decades. Nymphs emerge in May and June, becoming adult in late July. Adults can occasionally survive through to November.

Decticus verrucivorus (Linnaeus, 1758)

Wart-biter

Legend (map key):
- ■ 1990 onwards
- ● 1970-89
- ○ pre 1970
- ✕ introduction
- ▲ re-introduction

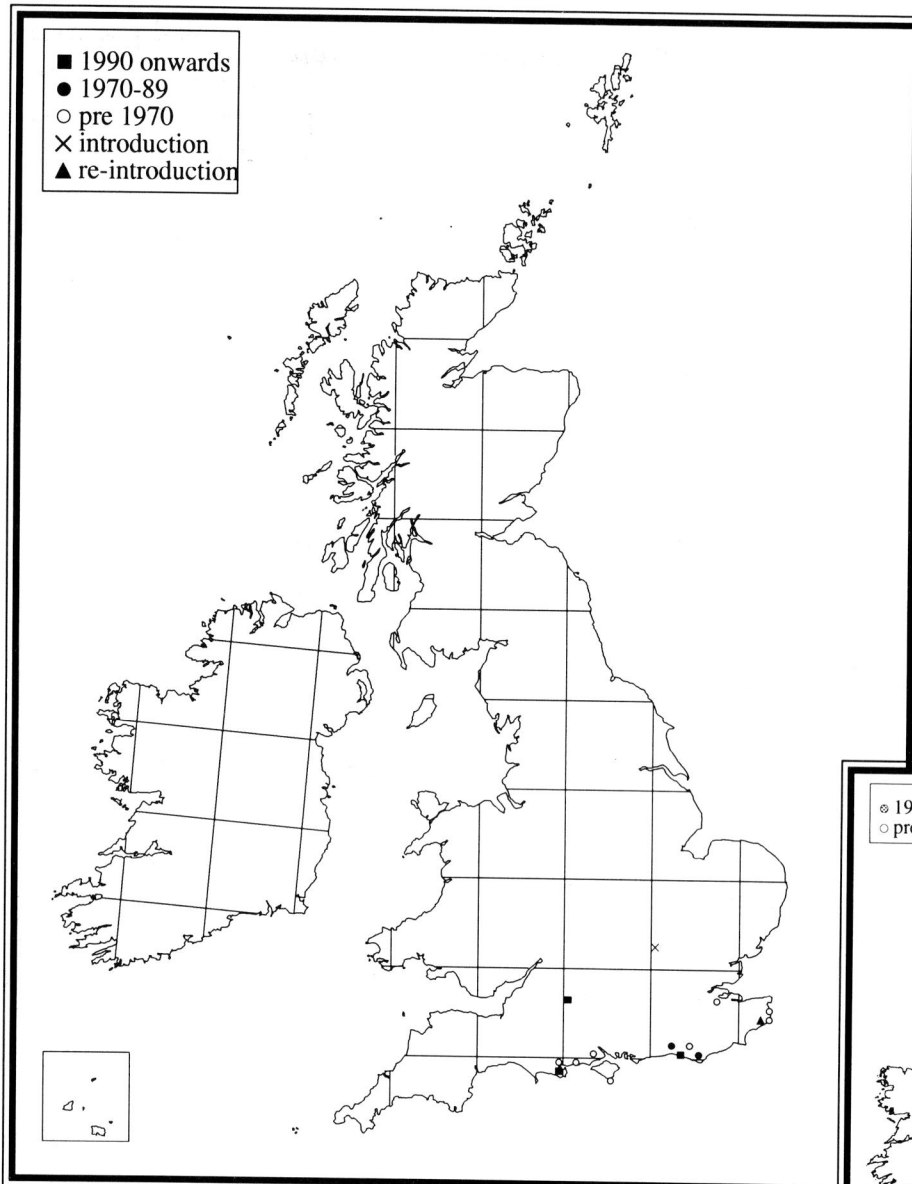

Marshall and Haes (1988) Description and species account pp82–84, Plate 1.

GB protection/threat status WCA Schedule 5, RDB2 (Vulnerable), Species Recovery Programme (under which re-establishment is on-going, within the species' original natural range).

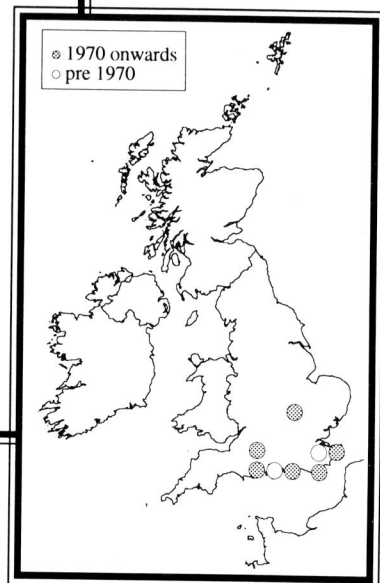

Inset map key:
- ⊗ 1970 onwards
- ○ pre 1970

Status

Recorded from a small number of isolated localities in southern England.

This large, flightless bush-cricket (31–37 mm) has an unmistakable frog-like appearance with powerful hind legs, speckled coloration and large, dark eyes. The wart-biter has been the subject of a detailed autecological study by Cherrill and Brown (1990a, b, 1991b). Colour variation has been recorded in nymphs and adults, including the typical green form and yellow/purple and grey forms (Cherrill & Brown 1991a). Variation in size also has been noted between specimens from downland and heathland sites (Cherrill & Brown 1992). Stridulation (a series of rapidly repeated clicks) is loud and distinctive, but males also produce a short stridulation similar to that of *Pholidoptera griseoaptera*. It occurs in very low numbers – the population at the best site in England was estimated to be 290 in 1987 and 190 in 1988 (Cherrill & Brown 1990a), but has increased in recent years due to appropriate site management. Its habitat requirements are complex and exacting – a mosaic of bare ground and long and short turf – so that few south-facing grassland or heathland sites are suitable for its survival. Oviposition occurs in bare soil or sparse grass (Cherrill, Shaughnessy & Brown 1991) and eggs remain dormant for at least two years. Egg hatch occurs in mid-April and adults emerge in July, but they rarely survive beyond mid-October.

Pholidoptera griseoaptera (De Geer, 1773)　　　**Dark bush-cricket**

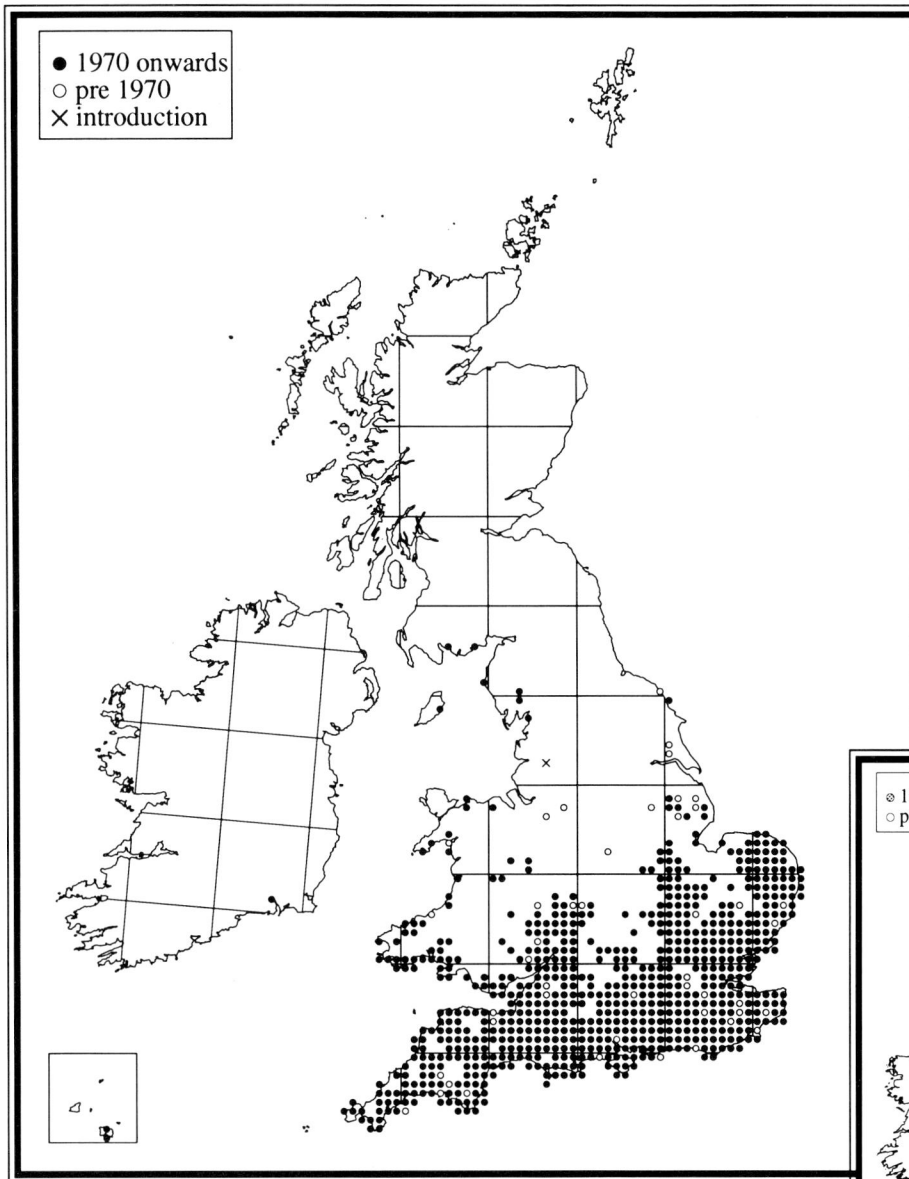

- ● 1970 onwards
- ○ pre 1970
- × introduction

Marshall and Haes (1988) Description and species account pp84–85, Plate 2.

IOM protection/ threat status Wildlife Act 1990 Schedule 5

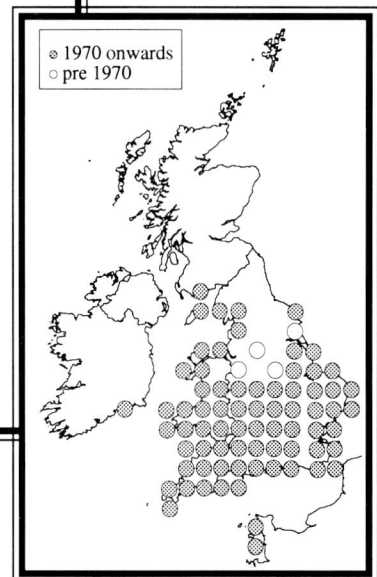

- ⊘ 1970 onwards
- ○ pre 1970

Status

Recorded from Britain, Jersey, Isle of Man and Ireland, but common only in the southern half of England. Northern and western populations are mainly coastal and/or near ancient woodlands, suggesting that they are probably natural and not accidental introductions.

This sturdy, dark brown species with yellow underside (13–20 mm) is almost wingless. Adults tend to remain well concealed in coarse vegetation and low bushes, but early instar nymphs frequently sit on leaves in full sunshine. Stridulation is a brief, penetrating *chip*, but as the species often occurs in some numbers and the sound carries well, an impressive chorus of single calls is not uncommon. The call is often the best way of locating a colony, in the afternoon, evening and even long after dusk. Wasteland, bramble thickets, old hedges, woodland edges and rides, thickets on sea

cliffs and scrub on the edges of saltmarshes and dunes are all likely sites for this species. Even wide, unmown, roadside verges and scrubby ditches in arable farmland can support small populations. There has been an increase in the geographical range of this species in parts of the Midlands and East Anglia in the last 50 years. Nymphs, which superficially resemble spiders, appear at the end of April and the adults, which emerge in late June or early July, can survive through to late November or early December, especially if there are no frosts.

Platycleis albopunctata (Goeze, 1778)

Grey bush-cricket

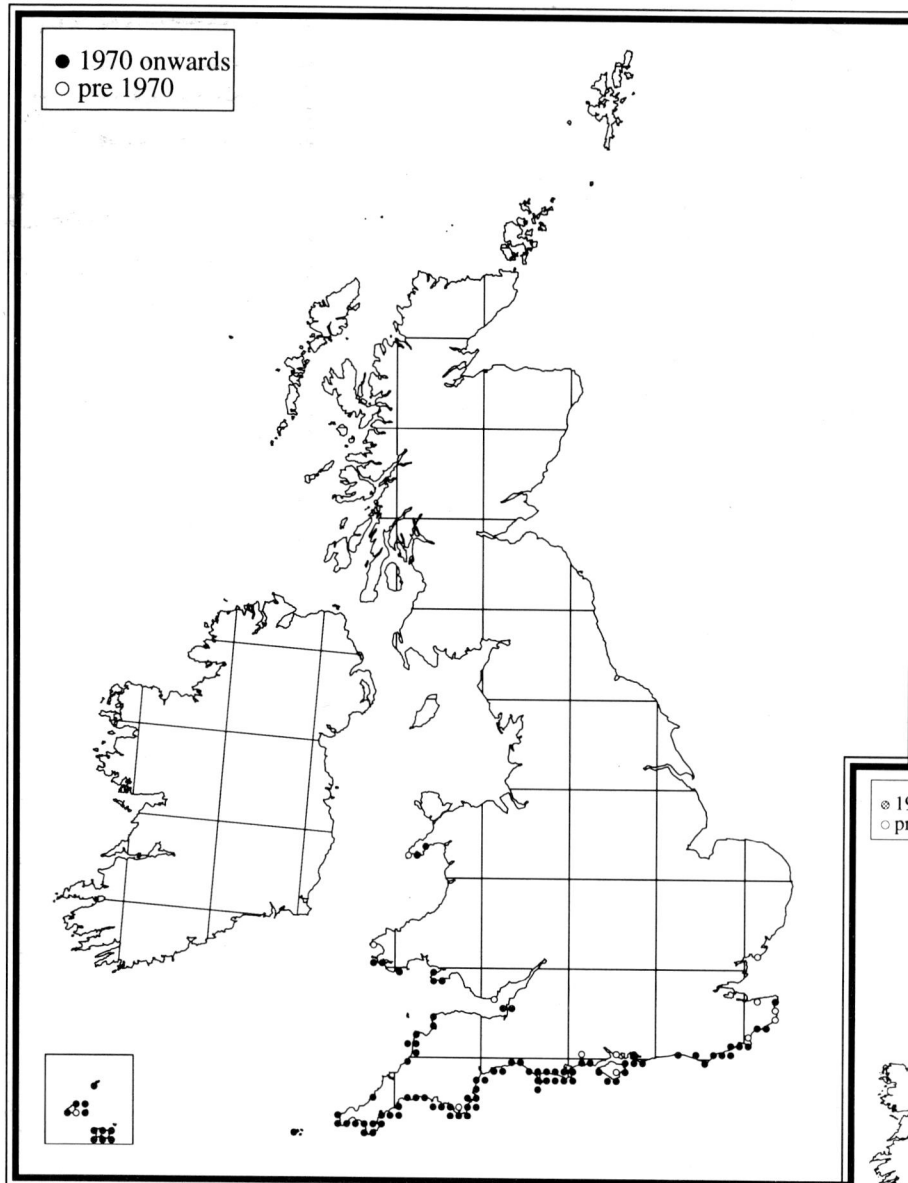

Legend:
- 1970 onwards
- ○ pre 1970

Marshall and Haes (1988) Description and species account pp85–86, Plate 2.

GB protection/threat status Nationally Scarce (B).

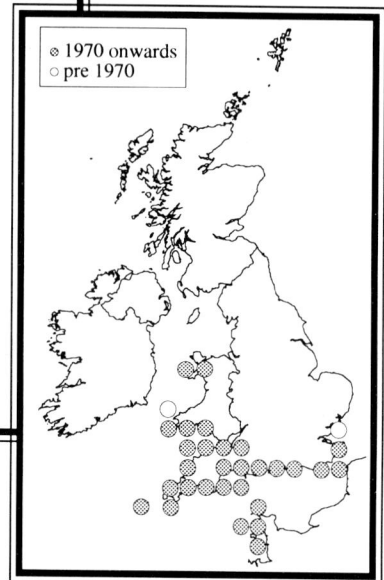

Status

Native on the south coasts of England and Wales and also occurs on all the major Channel Islands.

This large species (20–28 mm) is usually greyish brown, but coloration does vary. Both sexes are fully winged and readily fly short distances when disturbed. Stridulation is a soft, rapidly repeated, high-pitched chirp. It is a strictly coastal species in Britain, occurring among coarse grasses and rough vegetation on sand dunes, shingle beaches and south-facing sea cliffs, but has not been recorded from saltmarshes. The underlying rock type of sea cliffs does not seem to influence its occurrence. On the Continent, it is not restricted to coastal sites. Nymphs, which hatch in May, are often partly green in colour, and are so well camouflaged that they are easily overlooked. Nymphs may also be confused with those of *Metrioptera brachyptera*. Adults appear in July and rarely survive beyond the end of October.

Metrioptera brachyptera (Linnaeus, 1761)   **Bog bush-cricket**

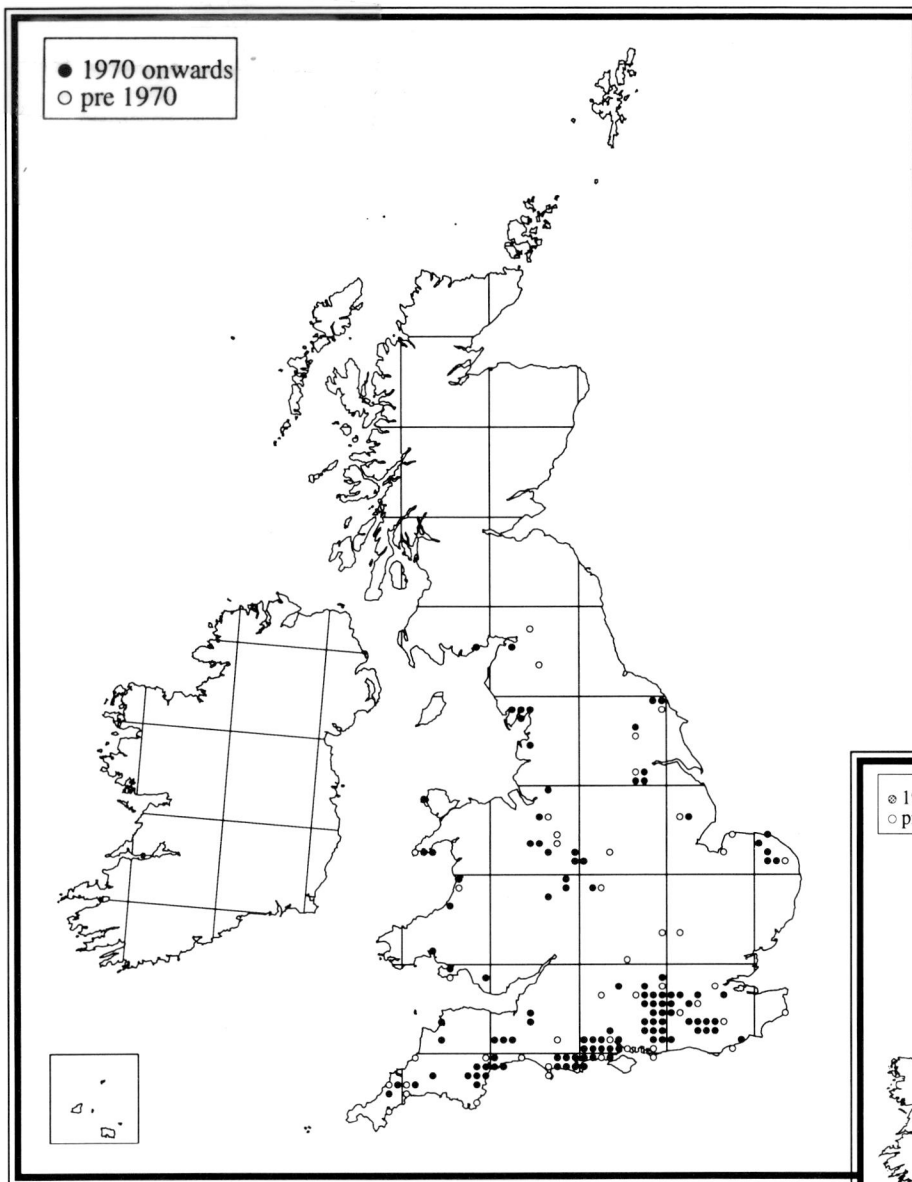

- ● 1970 onwards
- ○ pre 1970

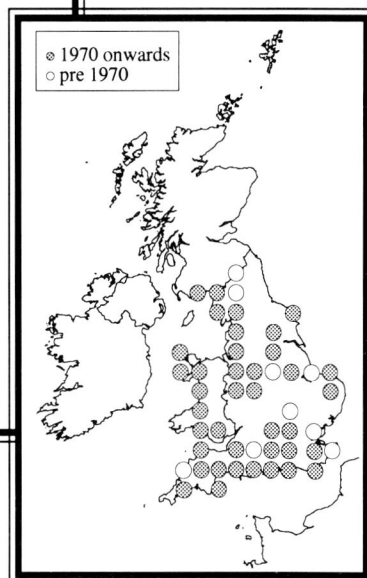

- ⊗ 1970 onwards
- ○ pre 1970

Marshall and Haes (1988) Description and species account pp86–87, Plate 2.

GB protection/threat status Nationally Scarce (B).

Status

Native in scattered heathy localities throughout England and Wales, but most common in the south of England (excluding Kent). Recently found at one site in south-west Scotland.

This medium-sized (11–21 mm), green and brown bush-cricket occurs in two colour forms, though the abdomen is bright green ventrally. Both sexes are normally brachypterous having only vestigial hindwings and short forewings, but there is a very rare fully-winged form, *marginata* (Thunberg), with dark brownish/black wings. Stridulation is a soft but shrill buzz, rather like a rapidly ticking watch, repeated several times, which is more easily heard in hot weather when several males stridulate together. It is a characteristic species of lowland heaths and clearings in damp heathy woodland, typically associated with the presence of cross-leaved heath (*Erica tetralix*) and purple moor-grass (*Molinia caerulea*). Nymphs hatch in May and June, and adults develop from July onwards, surviving in mild autumns until early November.

Metrioptera roeselii (Hagenbach, 1822)　　　　**Roesel's bush-cricket**

▲ colonisation since 1990
● 1970 onwards
○ pre 1970

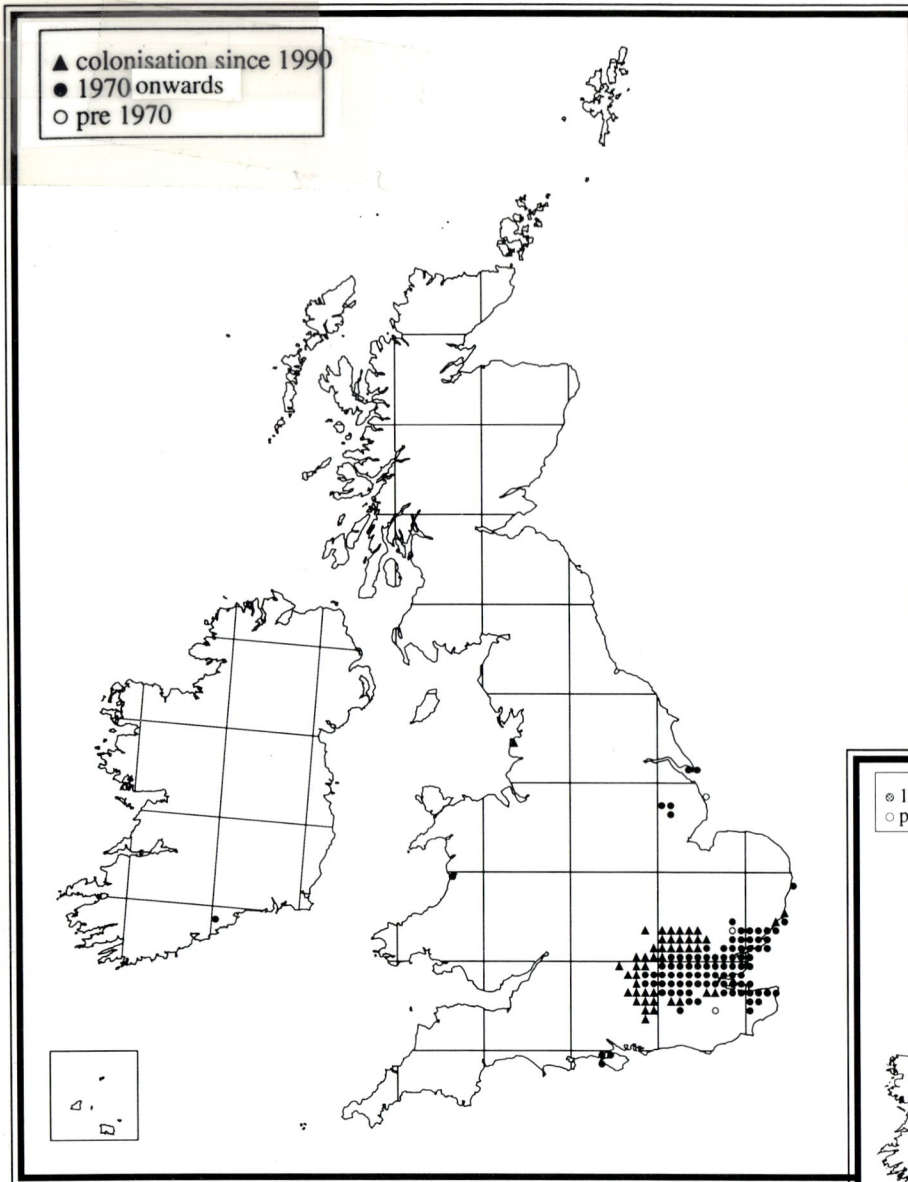

Marshall and Haes (1988) Description and species account pp87–89, Plate 2.

GB protection/threat status Nationally Scarce (B)

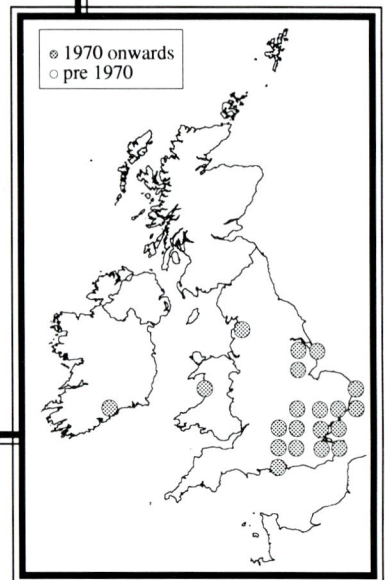

⊘ 1970 onwards
○ pre 1970

Status

Formerly coastal/estuarine. Now increasingly common and spreading in south-east England. Also recorded in Lincolnshire and as isolated coastal populations from Hampshire, Yorkshire, Lancashire, mid-Wales and southern Ireland.

A medium-sized (13–26 mm), normally dark brown and yellow species, sometimes with a greenish tinge; more rarely green forms occur. Both sexes are normally brachypterous, but a fully winged form, *diluta* (Charpentier), is sometimes frequent in hot summers. Stridulation is a penetrating and continuous high-pitched buzzing, like the crackle of overhead electricity cables, often heard late into the evening in hot weather. Originally recorded mainly from the landward side of saltmarshes and dunes on the North Sea coast, in the last 50 years it has spread westwards to urban wasteland, road and railway verges and recently to agricultural set-aside land, often in very large numbers. This range expansion has been very considerable since 1985 and appears to be continuing. It is usually found in coarse vegetation in warm and sunny locations at low altitudes, up to about 100 m on downland. Nymphs emerge in late May and June, becoming adult during late July. Specimens can be found through to late October in favourable years.

Conocephalus discolor (Thunberg, 1815)   **Long-winged cone-head**

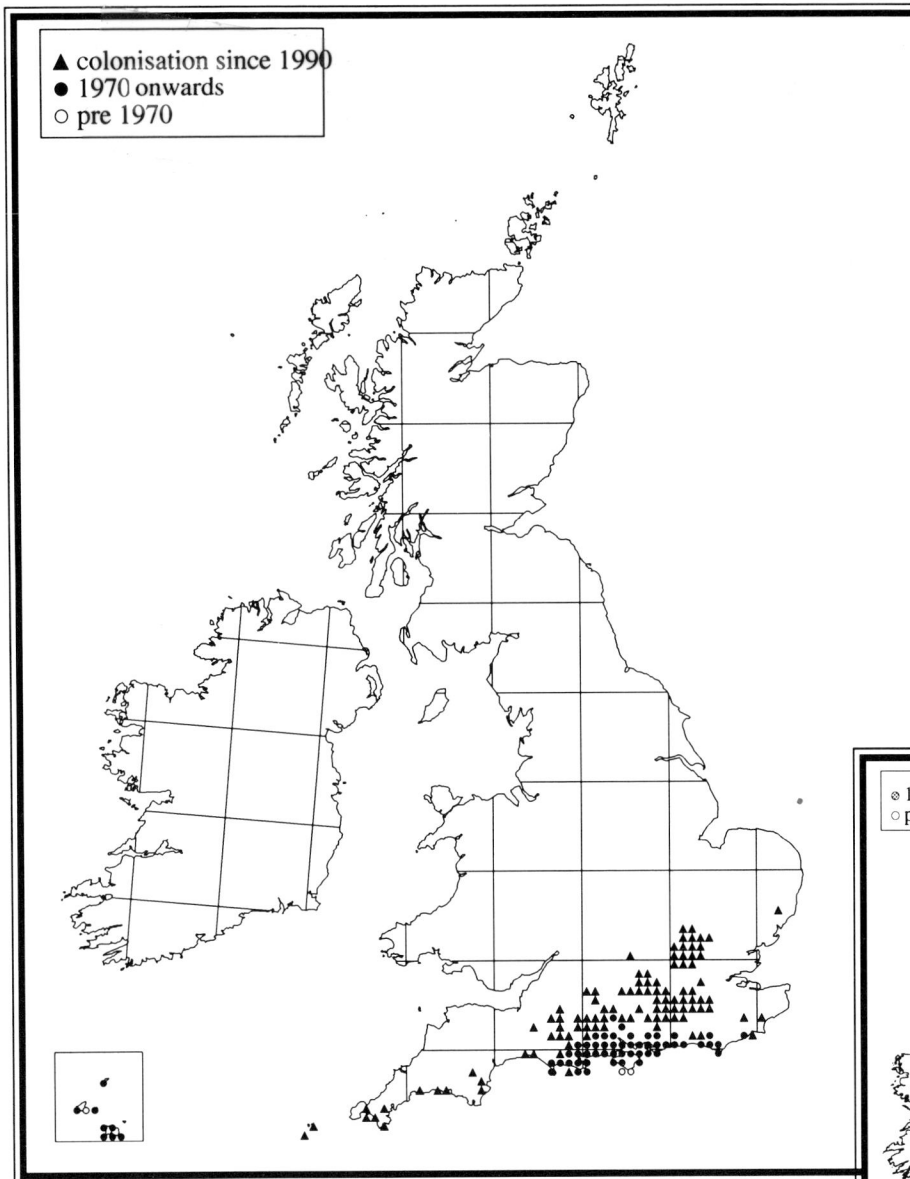

▲ colonisation since 1990
● 1970 onwards
○ pre 1970

⊘ 1970 onwards
○ pre 1970

Marshall and Haes (1988) Description and species account pp89–90, Plate 3.

GB protection/threat status Nationally Scarce (A).

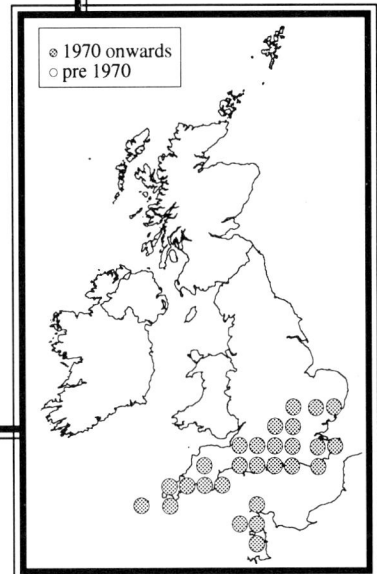

Status

Originally localised and uncommon along the south coast, it has spread rapidly in southern England since about 1970. Common on all the larger Channel Islands.

This slender, green bush-cricket (16–22 mm) has a brown dorsal stripe. Both sexes are fully winged, but there is an extra-macropterous form with much longer wings. Stridulation is a faint, prolonged hissing sound, which is at such a high frequency as to be inaudible to many people. It occurs in coarse vegetation in warm places in a wide range of habitats – ungrazed downland, urban wasteland, coastal reedbeds, dry heaths and even bogs. The spread of this species has been rapid and sustained for 20 years, increasing noticeably in hot summers, but individual populations fluctuate in numbers from one year to another. Nymphs emerge in May and June and mature to adults in August, which often survive until November in mild autumns.

Conocephalus dorsalis (Latreille, 1804)   **Short-winged cone-head**

Marshall and Haes (1988) Description and species account pp90–91, Plate 3.

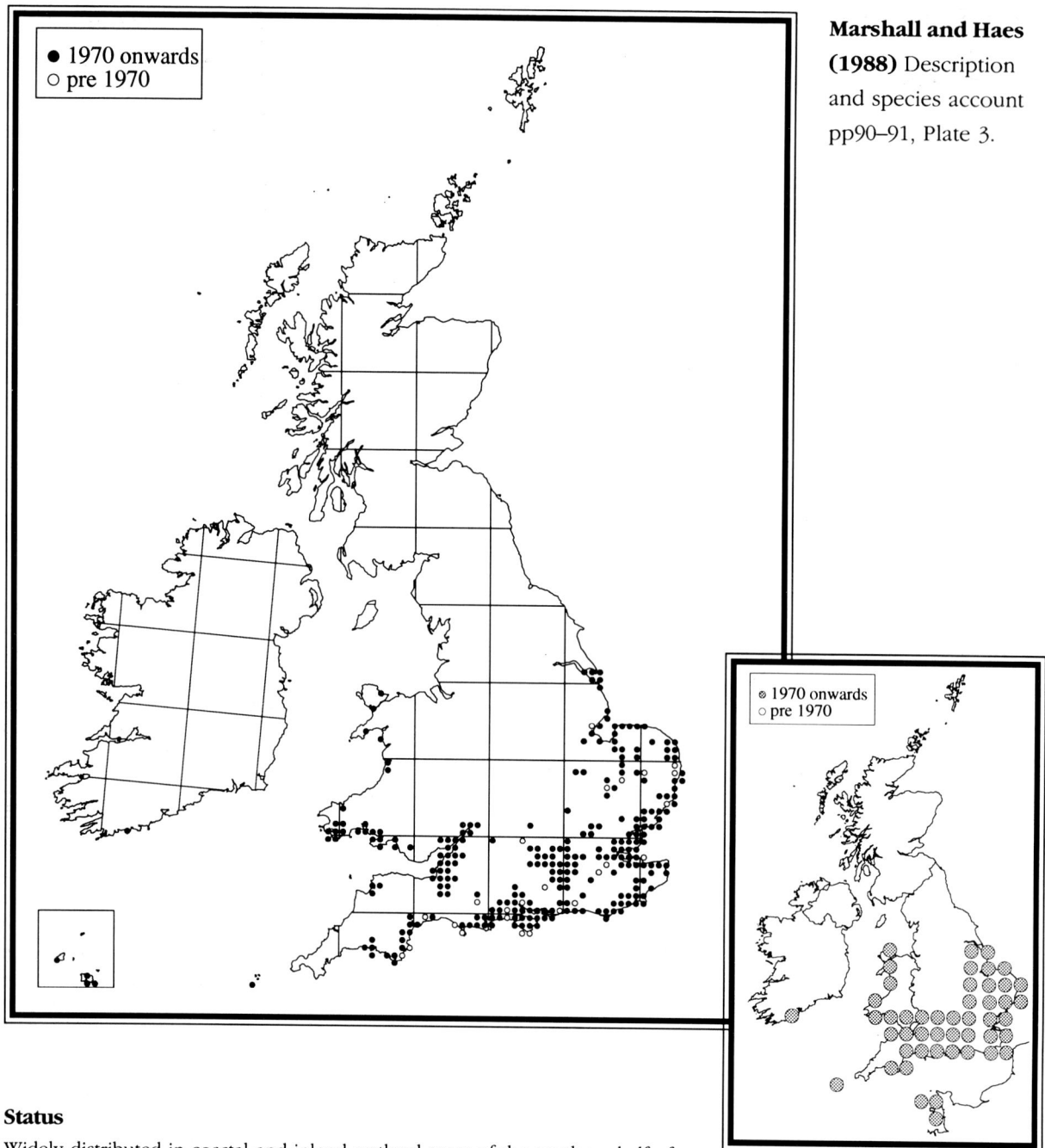

● 1970 onwards
○ pre 1970

⊘ 1970 onwards
○ pre 1970

Status

Widely distributed in coastal and inland wetland areas of the southern half of England and Wales and also recorded from single sites in Guernsey, Jersey and south-west Ireland.

This small, green bush-cricket also has a brown dorsal stripe, but is generally more compact (11–18 mm) than *C. discolor*. Both sexes are normally brachypterous, but there is a macropterous form (*burri* Ebner) with both pairs of wings fully developed, which is found mainly in hot summers. Stridulation is two alternating sounds – a faint, prolonged hiss or buzz, rather like that of *C. discolor*, and a rapid ticking; however, it is faint and often difficult to hear. This species occurs in two distinct habitats: coastally on saltmarshes and sand dunes, particularly associated with maritime rushes and grasses, and inland on lowland bogs, fens, reedbeds, river floodplains and by lakes and pools. Unlike *C. discolor*, it rarely occurs in large numbers. Well over half the known sites for this species have been found for the first time during the period of the survey, but it was almost certainly under-recorded in some areas until recent decades. It is, however, certainly spreading in the vicinity of the Thames valley and in Hertfordshire. Nymphs emerge in May and June and mature to adults in July and August, but these rarely survive beyond mid-October.

Leptophyes punctatissima (Bosc, 1792)   **Speckled bush-cricket**

1970 onwards
pre 1970

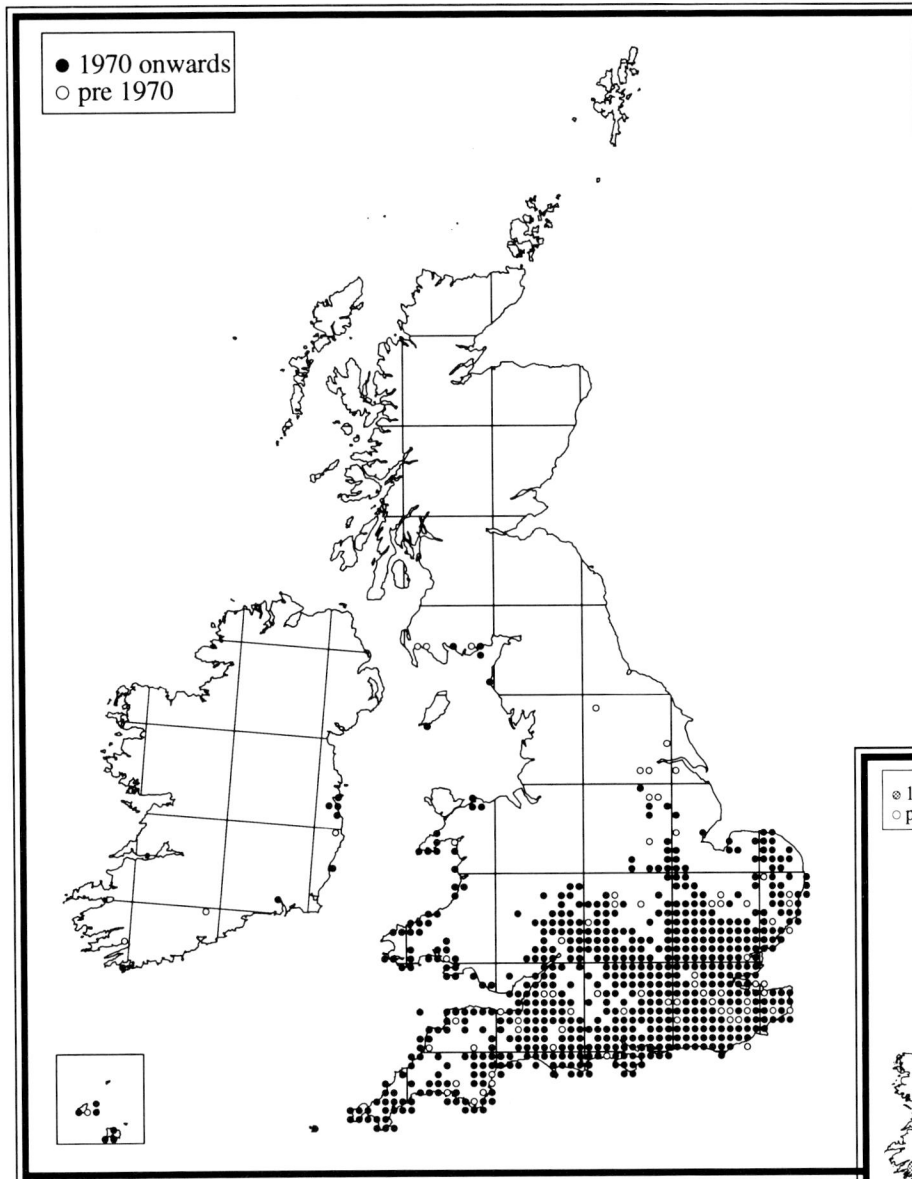

Marshall and Haes (1988) Description and species account pp92–93, Plate 3.

IOM protection/ threat status Wildlife Act 1990 Schedule 5.

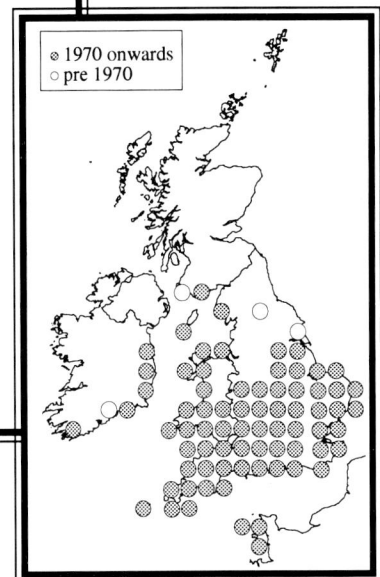

Status

Recorded from Britain and Ireland, but most widely distributed in the southern half of England, north to the Humber. The northern and western populations, in Wales, south-west Scotland and the south and east of Ireland, are mainly near the coast. It has been recorded from several coastal islands.

A small (9–18 mm), distinctly plump and hunch-backed, green bush-cricket which is suffused with minute dark spots. Both sexes are brachypterous, with the hindwings absent. Stridulation by both sexes is a high-pitched chirp repeated regularly and almost inaudible to most humans. It occurs in rough vegetation and scrub in a range of habitats, including woods, hedges, waste ground and gardens. It also occurs in saltmarshes, reedbeds and fens where males might be mistaken for *C. dorsalis*. It is certainly the bush-cricket which is most frequently *seen* in southern Britain and will come to lighted windows at night in early autumn. Nymphs appear in May and adults in July, surviving to late October or early November in mild autumns.

Acheta domesticus (Linnaeus, 1758)

House cricket

Marshall and Haes (1988) Description and species account pp94–95, Plate 3.

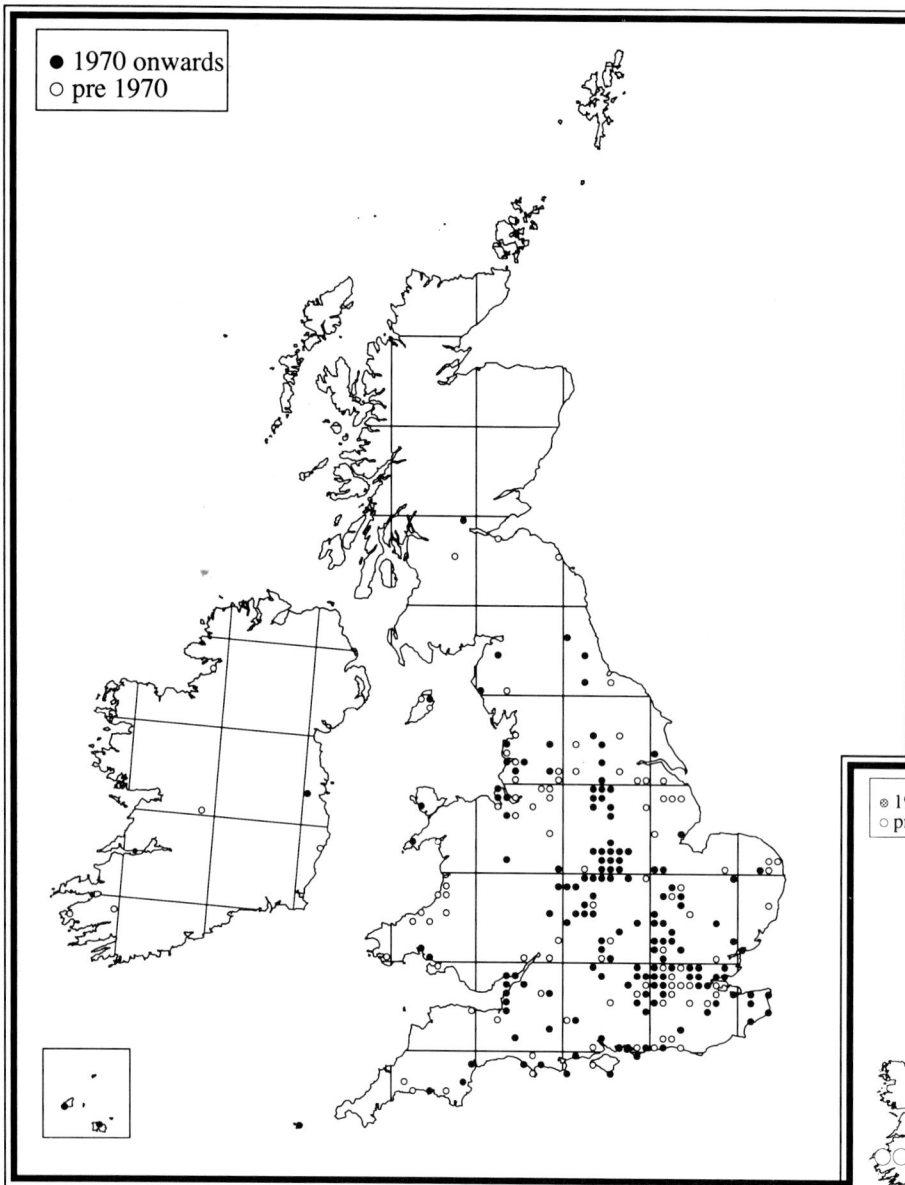

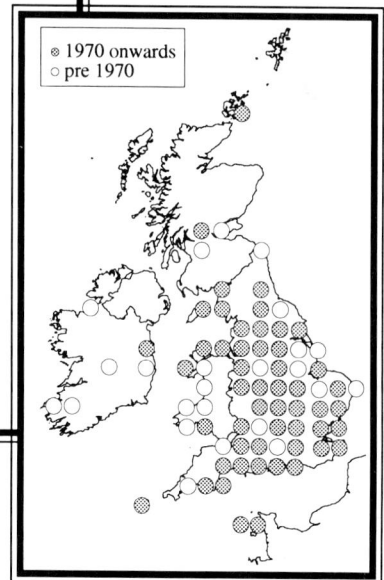

Legend (main map): ● 1970 onwards ○ pre 1970

Legend (inset map): ⊗ 1970 onwards ○ pre 1970

Status

Not native, but widely distributed in England, with only scattered, mainly old, records from Wales, Scotland and Ireland. Probably declining, but now widely sold as live food for pets (eg exotic reptiles) and apt to escape.

This medium-sized (14–20 mm), greyish-brown cricket is widely held to have originated in North Africa and the Middle East, but has clearly been established in Britain (and probably in Ireland) for many centuries. It is more often heard (a distinctive, regular chirp) than seen, being active only at night. Both sexes are fully winged. It was formerly a common and familiar insect in domestic houses, bakehouses and breweries (see, for example, White 1789). Over the last century, it has become more restricted to large, continuously heated, institutional buildings, such as hospitals, hotels, factories and a few laboratories, where it occupies central heating and other large circulatory ducting systems. These populations are usually subject to regular pest control treatment. It has been recorded at some large domestic refuse tips, but there have been a few other outdoor records of singing males in hot summers. As it lives in artificial conditions, there is no seasonality or diapause.

Gryllus campestris (Linnaeus, 1758)

Field cricket

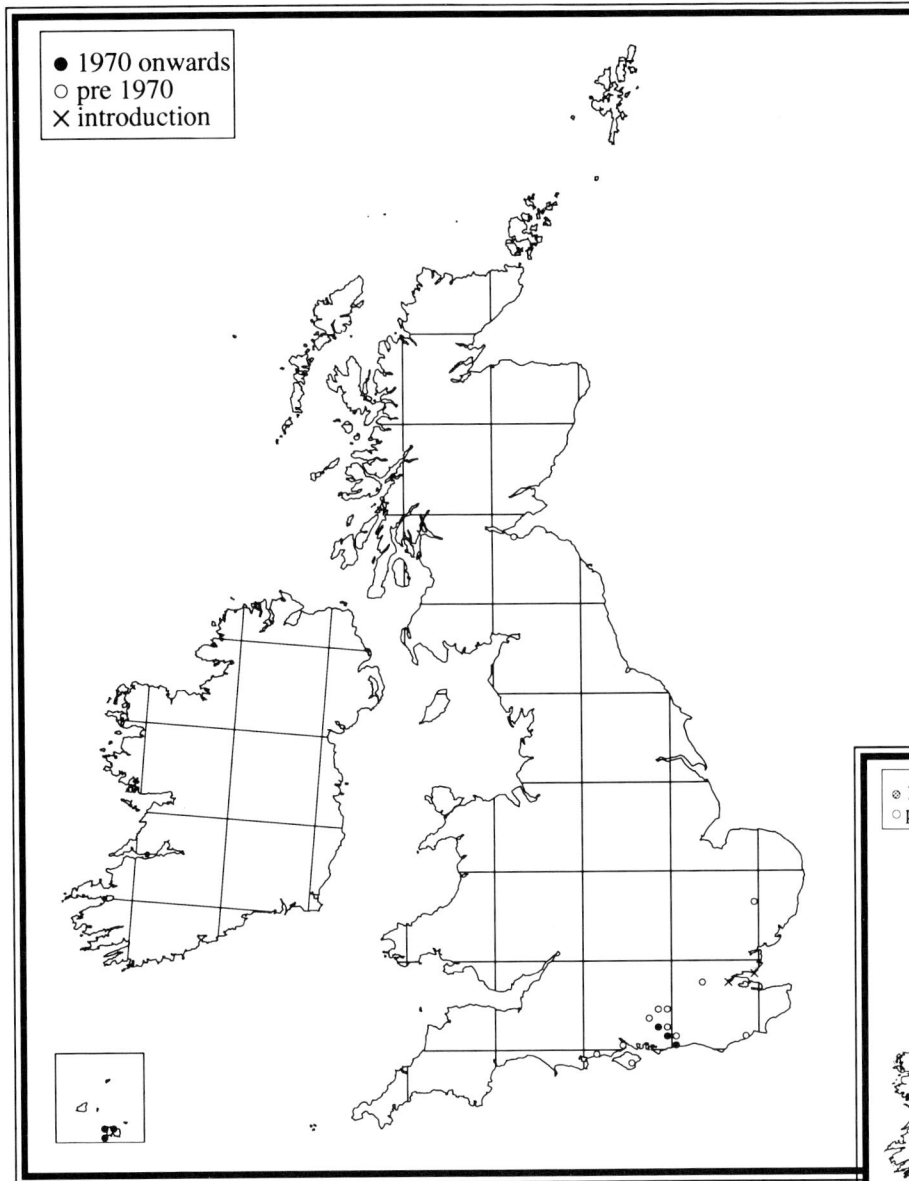

- ● 1970 onwards
- ○ pre 1970
- × introduction

Marshall and Haes (1988) Description and species account pp95–96, Plate 3.

GB protection/threat status WCA Schedule 5, RDB1 (Endangered), Species Recovery Programme (under which four re-establishments have been initiated and more are planned, all within its original natural range).

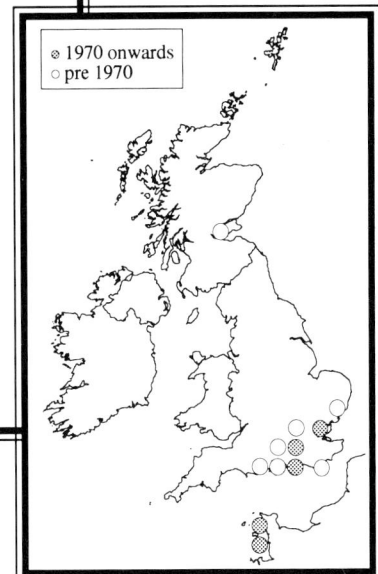

- ⊘ 1970 onwards
- ○ pre 1970

Status

Native in England and probably always very localised to eastern Hampshire, the Isle of Wight, Surrey and West Sussex. By 1994 restricted to one original site in West Sussex. Marked decline this century, but some early records may be the result of confusion with *Acheta domesticus*, which can become temporarily established outdoors in hot summers.

This is a large (19-23 mm) and distinctive, shiny, black cricket. Both sexes have forewings, with yellow basal patches, but only vestigial hindwings. Males have a loud song made up of short chirps repeated several times a second. Because of the colonial nature of populations of this species, choruses of males can sometimes be heard at distances of up to 100 m. There is a quieter courtship song. In England, field crickets require areas of south-facing short turf on sandy or chalk soils which have small areas of bare ground. Many of the known former sites have been lost to development, recreational use, arable or the development of scrub or woodland. Older nymphs and adults live in distinctive burrows with horizontal entrances, overwintering there as well-developed nymphs. Eggs are laid from May to July, and the adults die by August. Although primarily herbivorous, this species will also scavenge dead insects. Recent research has shown that there is considerable interaction between individuals in a population and mobility between burrows.

Nemobius sylvestris (Bosc, 1792)

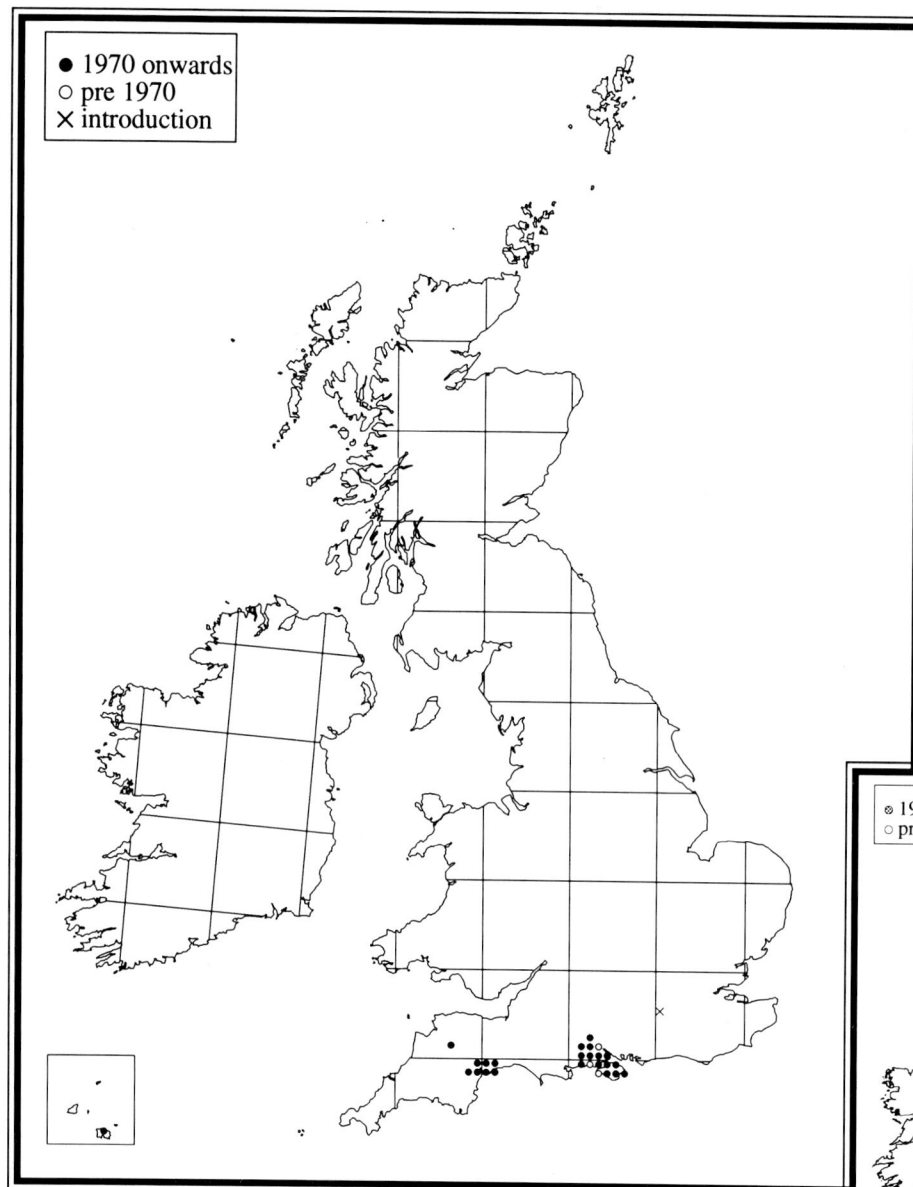

Marshall and Haes (1988) Description and species account pp96–97, Plate 4.

GB protection/threat status Nationally Scarce (A).

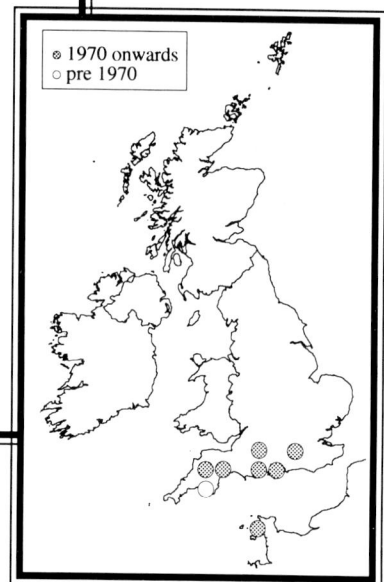

- ● 1970 onwards
- ○ pre 1970
- ✕ introduction

⊗ 1970 onwards
○ pre 1970

Status

Status in Britain uncertain, probably native in wooded areas in the south of England, including the Isle of Wight and in Jersey. Range has possibly declined, but now fairly stable.

This small (7–11 mm), dark-brown cricket with pale markings lives mainly in deep leaf litter, but when disturbed is very active and sprightly. Both sexes are entirely flightless with only reduced forewings (longer in males). It is usually associated with warm clearings and the sunny edges of woods or areas of scrub, but it also sometimes occurs in stone walls, earth banks and clay sea cliffs near woods or scrub. It probably feeds mainly on dead leaves and fungi. It often occurs in large numbers, with both nymphs and adults present at the same time, because of the unusual two-year life cycle (Brown 1978). The normally low purring song of a solitary male is difficult to hear, but a chorus of males can be a distinctive feature of the few locations where they occur and can be heard (day and night) from July to November. The population in Surrey may have originated from an accidental introduction in the early 1960s with plants from the New Forest, but, if so, it has since spread. A deliberate introduction of nymphs from the New Forest and adults from Surrey, to a park in west London, was reported in 1996 (not mapped). There are unconfirmed records including from a wood in the Severn Valley, Worcestershire, in 1948, and from Derbyshire (Kevan 1961). Until the end of the 19th century, its only known locality was at Lyndhurst, in the New Forest, where it was originally found by J C Dale in 1820.

Pseudomogoplistes squamiger (Fischer, 1853)   **Scaly cricket**

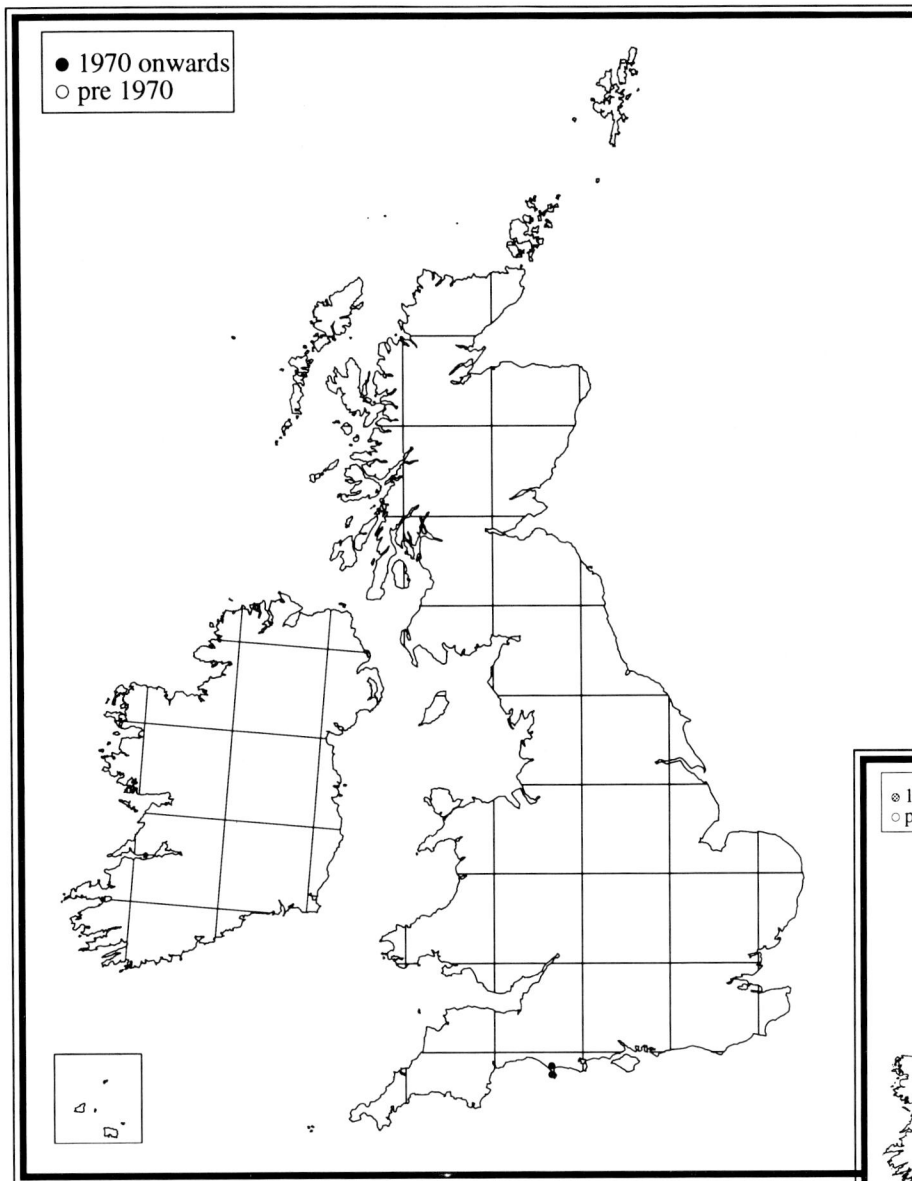

● 1970 onwards
○ pre 1970

**Marshall and Haes
(1988)** Description
and species account
pp97–98, Plate 4.

**GB protection/threat
status** RDB1
(Endangered).

⊗ 1970 onwards
○ pre 1970

Status

Probably introduced and successfully naturalised, occurring only at Chesil Beach,
Dorset. Population known since 1949.

This small cricket (8–13 mm) is greyish brown in
colour, with most of the body and legs covered with
minute scales. Adults are completely wingless and
therefore males cannot stridulate. It is restricted to the
eastern half of Chesil Beach on the shore of The Fleet,
a brackish lagoon. It probably lives by day among
shingle and under stones, appearing at night to
scavenge on the sea shore and in the intertidal zone of
the lagoon. Nymphs and one adult have been found as
early as April, but most adults have been recorded from
August to October. Eggs are laid in the autumn and
probably overwinter, with nymphs emerging in the
spring. It was thought to occur in very small numbers
at Chesil Beach until more than 90 dead adults were
found in a single discarded plastic carton in October
1992 (Timmins 1994a, b) and as a result of surveys in
1994. This species is distributed around the
Mediterranean coasts, the Algarve in Portugal, and
some Atlantic islands. Its occurrence at Chesil Beach
may be the result of an accidental introduction, but
other shingle-dwelling invertebrates with a mainly
Mediterranean distribution are known to occur
elsewhere on the coasts of southern England and
Wales.

Gryllotalpa gryllotalpa (Linnaeus, 1758)   **Mole cricket**

● 1970 onwards
○ pre 1970

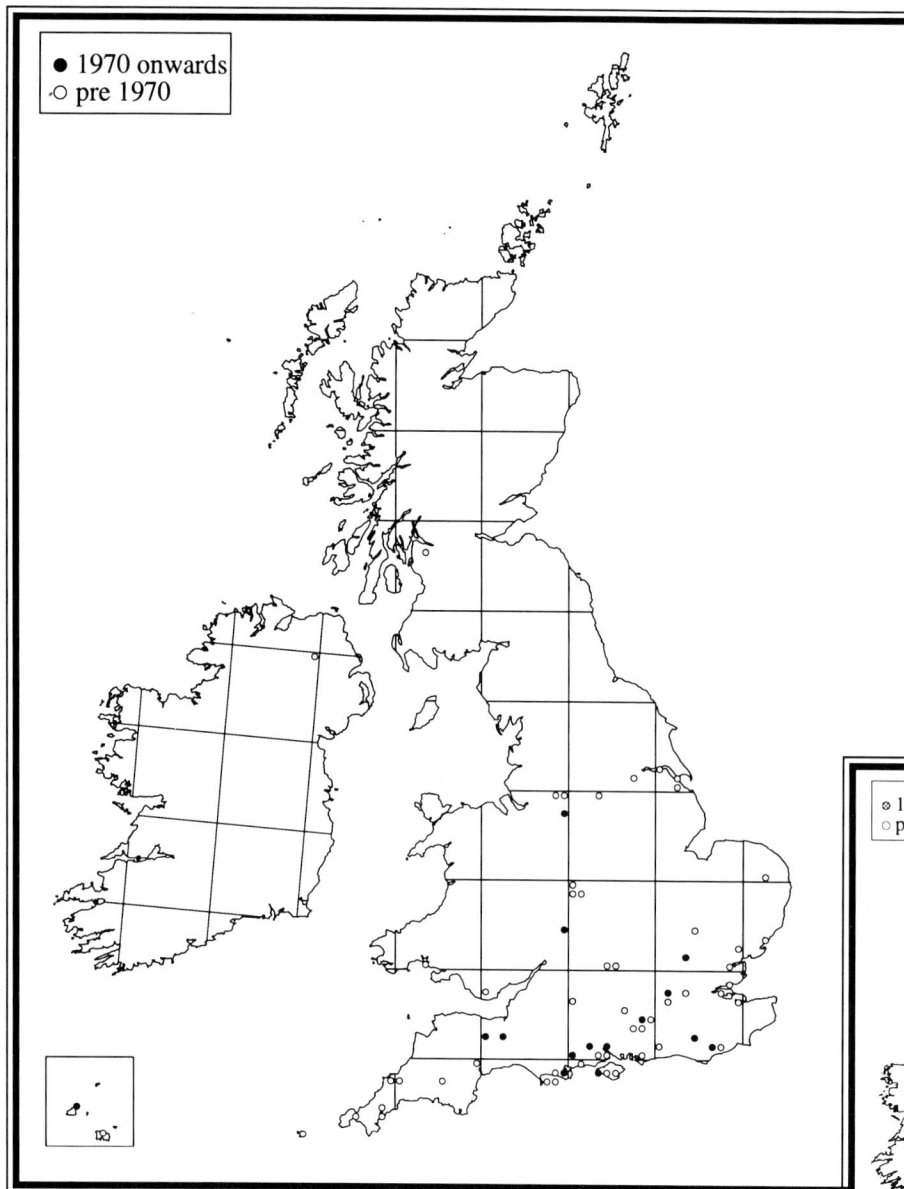

Marshall and Haes (1988) Description and species account pp98–99, Plate 4.

GB protection/threat status WCA Schedule 5, RDB1 (Endangered), Species Recovery Programme. Listed as a priority threatened species in *Biodiversity: the UK Steering Group Report* 1995

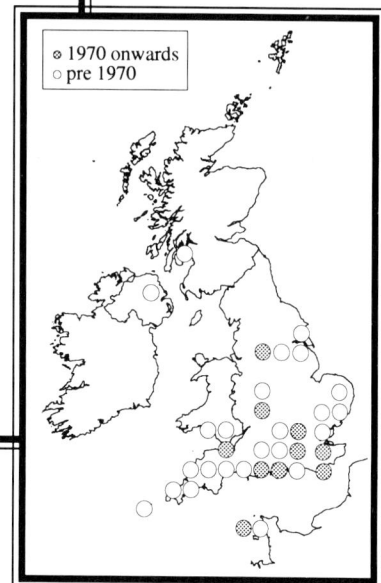

⊗ 1970 onwards
○ pre 1970

Status

Native and formerly widely distributed in England, Wales and the Channel Islands, but always restricted to areas of moist soils. Formerly known definitely from one locality in Scotland and possibly from a single locality in Northern Ireland. Although undoubtedly much reduced in frequency, this is a very secretive animal and it may well survive in a number of locations.

The mole cricket is large (35–46 mm), with an unmistakable shape. It is light chestnut-coloured and covered with velvet-like hairs; the forelegs are modified for digging – a truly spectacular insect. Although the forewings are short, both sexes are able to fly, but they do so only at night. The flight is noisy and clumsy. Male stridulation is a loud continuous purring, usually from burrow entrances on warm nights in spring and early summer. Stridulation is more easily confused with the song of a nightjar or a grasshopper warbler than any other orthopteran. It is likely, however, that individual males sing only once or on few occasions. The life cycle typically takes three years in Britain. The mole cricket was once a widespread and familiar insect, with many local names, and was regarded as a pest in gardens, although it feeds mainly on larvae and earthworms. It was regarded as an agricultural pest in parts of Guernsey until the 1980s and is still locally present in large numbers. It has been affected by loss of its habitat – mainly damp edges of wetlands and local seepages. Despite its size, it is an elusive species, spending most of its time underground, burrowing to considerable depths in loose, damp soil. Although the scattered, isolated records of this species typical of the past 50 years have been regarded as casuals, it is now thought that these may well represent genuinely native insects. Many of these records follow a run of warm summers, for example the group of records in 1988-89, and may reflect the build-up of local populations.

Tetrix Ground-hoppers

The size range of all three *Tetrix* species is about 8–14 mm and their appearance is superficially similar, each with several colour and pattern forms. They look like small grasshoppers, but with wide 'shoulders' and a narrow, tapering abdomen hidden beneath an extended pronotum. Because of their similarity, identification of *Tetrix* species in the field should be done with caution, especially as they do not stridulate. Mature adults of *T. ceperoi* and *T. subulata* are capable of flight and nymphs and adults of all three species are able to swim. Late instar nymphs and immature adults overwinter and then mature in the following spring. Eggs are laid from late April onwards, with the adults surviving only until July. Nymphs appear from May to July, with the new generation adults appearing from August onwards, but they do not mature immediately. *Tetrix* species are diurnal and feed on algae and mosses. The best time to observe them is late April to early June, when mature adults are present and courtship and mating can be observed; this is several months earlier than for grasshoppers. All three species have been under-recorded in the past and the two commoner species may still be somewhat under-recorded, especially in Ireland.

Tetrix ceperoi (Bolivar, 1887)

Cepero's ground-hopper

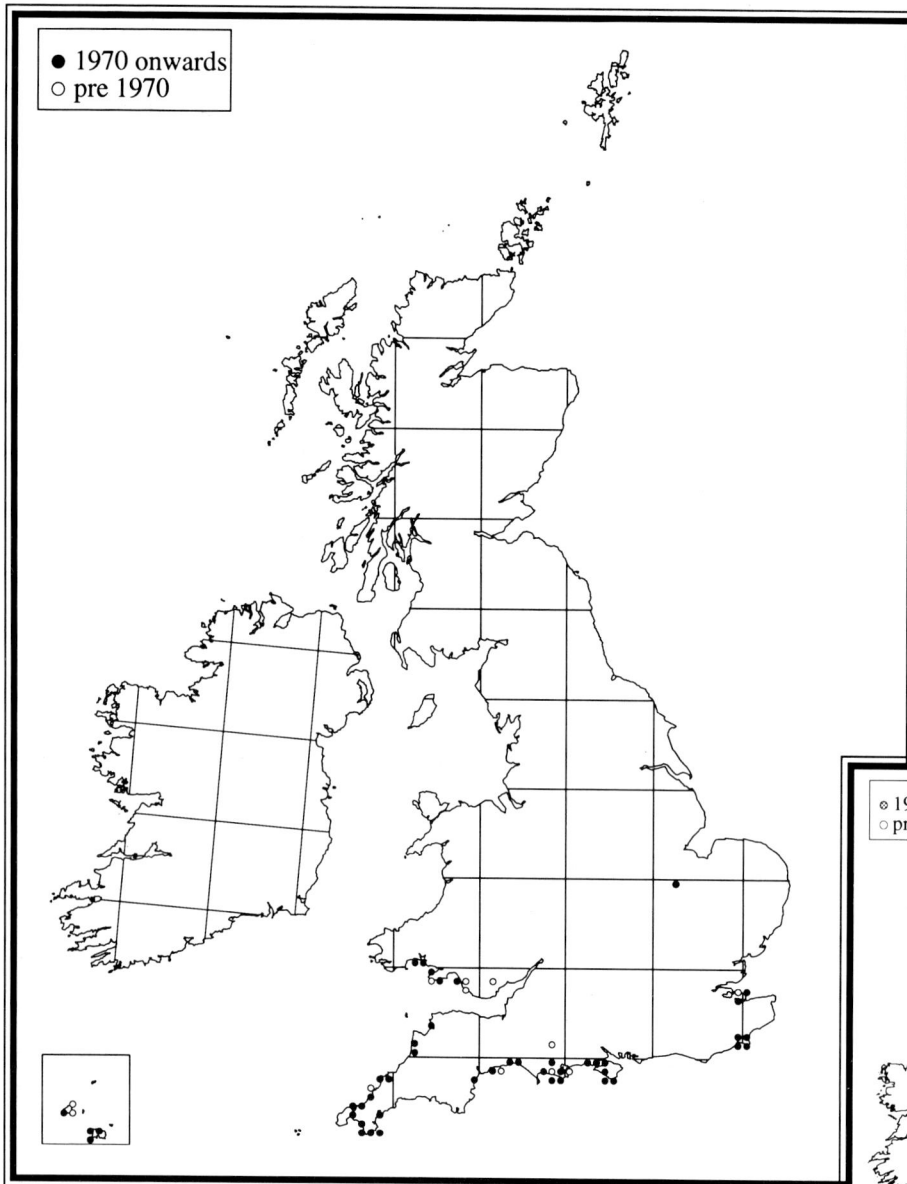

Marshall and Haes (1988) Description and species account pp100–101, Plate 4.

GB protection/threat status Nationally Scarce (A).

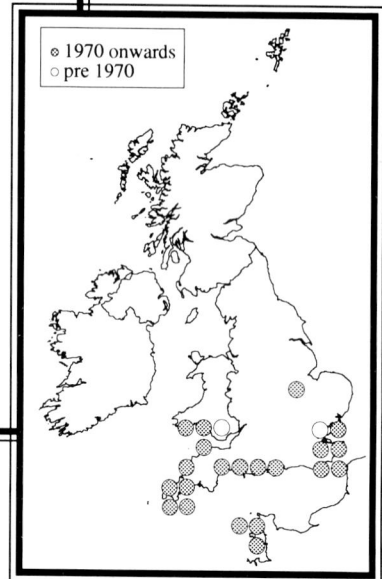

Status

Native and very locally distributed along the south coasts of England, Wales, Guernsey and Jersey. On the north-western edge of its range in Britain.

At favourable sites with plenty of open, sunny situations, this species can occur in large numbers. Many records are from coastal sites, but the actual habitats vary considerably – sand dunes and dune slacks, shingle banks, drainage dykes near saltmarshes, and seepages on sea cliffs. Inland localities include bare peat by streams and ponds in the New Forest and Dorset and on the Lizard peninsula. A record from Surlingham, Norfolk, was later found to be *T. subulata*, but the record from north Cambridgeshire has been confirmed. The next most northerly sites are around the sheltered bay formed by the estuaries of the rivers Taf, Towy and Gwendraeth, south of Carmarthen. Its restricted occurrence and geographical range are probably limited by the presence of favourable micro-climatic conditions. The species may occur in association with *T. subulata*.

Tetrix subulata (Linnaeus, 1758)

Slender ground-hopper

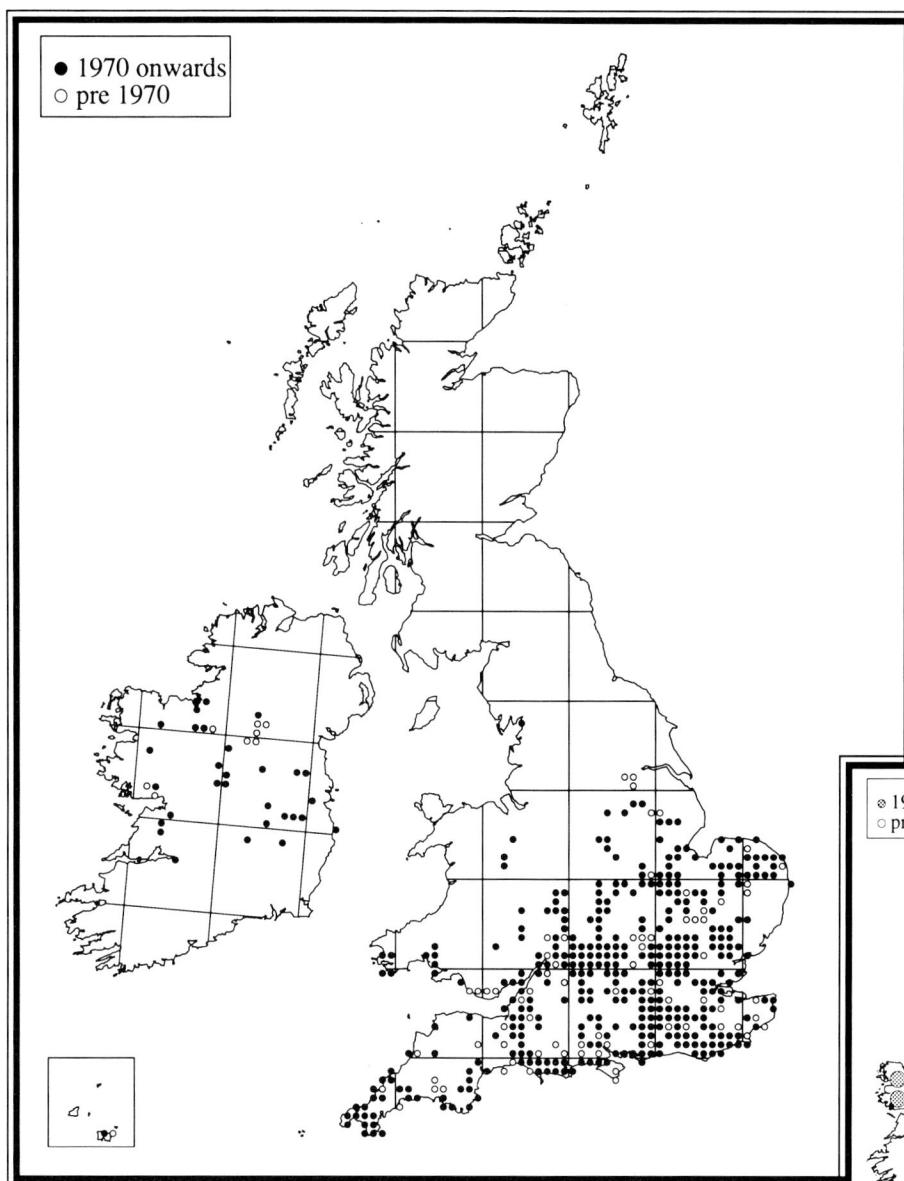

Marshall and Haes (1988) Description and species account pp101–102, Plate 4.

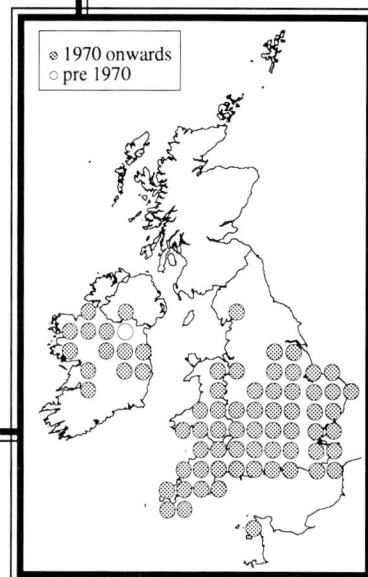

● 1970 onwards
○ pre 1970

⊘ 1970 onwards
○ pre 1970

Status

Native in Britain and Ireland. Widely distributed in southern and central England, more sparsely distributed elsewhere, including Jersey, but probably still under-recorded.

This species is similar in appearance to *T. ceperoi.* A short-winged form (f. *bifasciata* Herbst) may occur with a shorter pronotum and hindwings than normal. The species is characteristic of bare mud and short vegetation in damp, unshaded locations. It is particularly associated with base-rich or calcareous soils, such as dune slacks, limestone sea cliffs and floodplains and fens where the ground water is alkaline. Knowledge of the occurrence of this species in Wales has improved recently, but it still appears to be very localised. Bearing in mind the preference of this species for base-rich soils, it may be more widely distributed in the Welsh Marches than the present map suggests. The site of the most northerly record in England, at Silverdale Moss, Lancashire, may have been destroyed since it was discovered in 1987, but further surveys are needed in this area which is well known for the presence of northern outlying populations of several other invertebrates. It was formerly thought to be very localised in Ireland, but since about 1975 it has been found at many more sites, including several rich fens and a turlough. These finds have been largely the result of surveys by experienced orthopterists, and further surveys are needed.

Tetrix undulata (Sowerby, 1806)

Common ground-hopper

● 1970 onwards
○ pre 1970

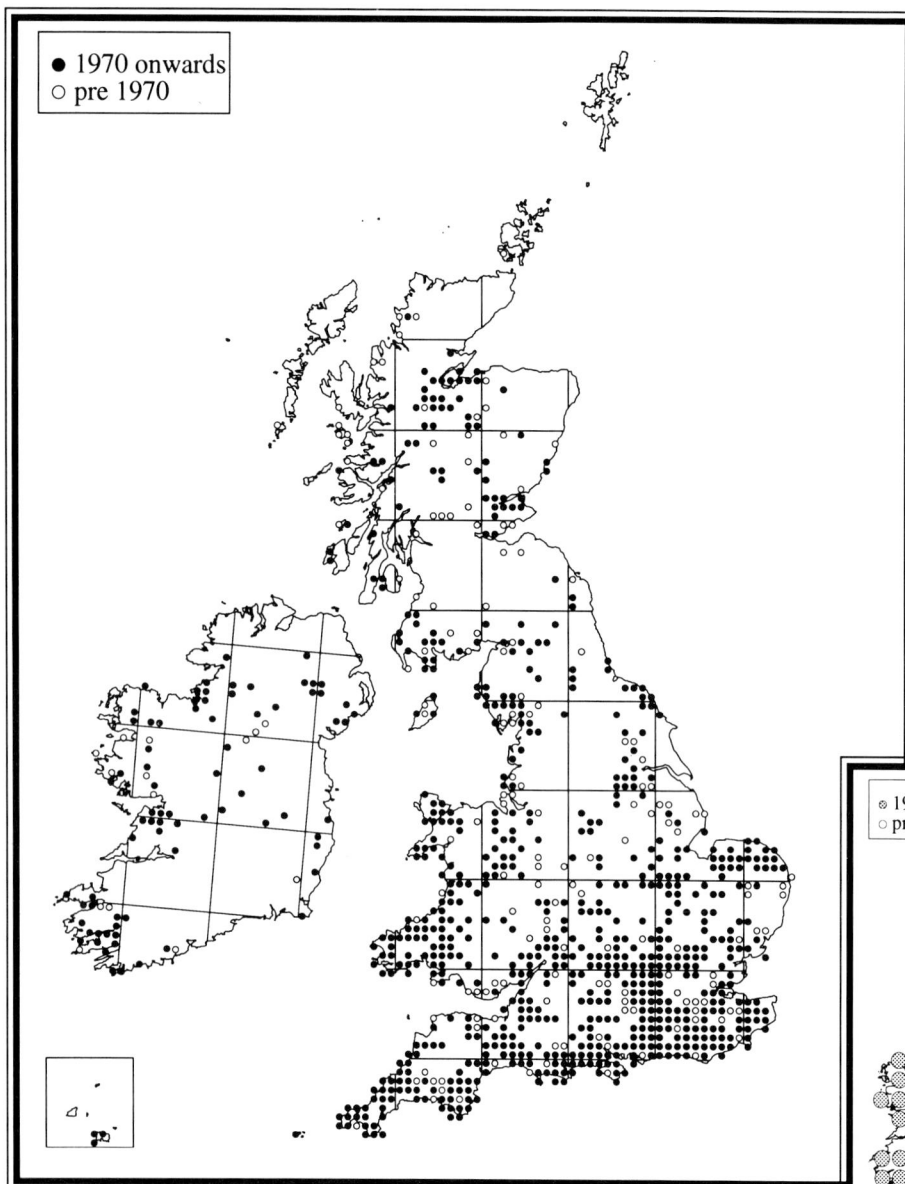

Marshall & Haes (1988) Description and species account pp103–104, Plate 4.

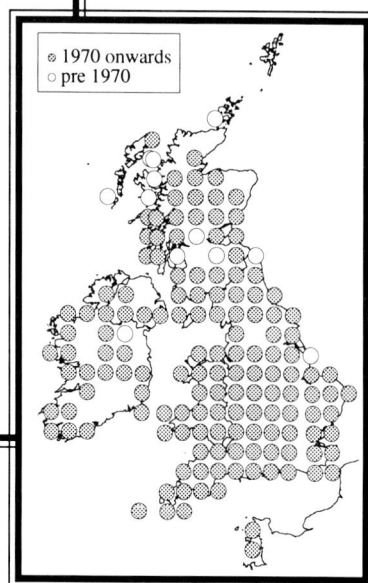

⊕ 1970 onwards
○ pre 1970

Status

Native in Britain and Ireland and widely distributed throughout. Although the majority of records are from southern and central England, it is probably still under-recorded throughout its range, which includes Jersey.

This species is the most robust-looking of our three species of ground-hopper. It requires an open habitat with bare ground and short vegetation, but, unlike the other two *Tetrix* species, it occurs in both wet and dry locations. The presence of mosses in the vegetation is also important. It can be found in suitable areas by sweeping vegetation and by sieving plant litter and can occur as adults and nymphs at any time of the year. It is most frequently found in damp places beside streams, in wet areas on heaths, in old quarries and alongside tracks in chalk and limestone areas. In woodlands and plantations, it can become very common for a few years in newly coppiced or clear-felled areas, until the vegetation shades out its preferred habitat. However, it seems able to persist in small numbers until suitable

conditions develop for the population to increase again. In Scotland, it has been recorded from several islands on the west coast and, in the Highlands, it has been recorded most frequently in association with relic areas of native pine forests. The very rare semi-macropterous form (f. *macroptera* Haij) has been recorded only in the Highlands, most recently in 1991 at Daviot (Easterness). In Ireland, this species has also been found in mountainous areas (Nephin Beg range (West Mayo) and Mourne Mountains (Down)) and some west coast islands, but the results of surveys in Sligo suggest that it may be more widespread than the present map suggests. As with the other two *Tetrix* species, *T. undulata* is probably overlooked by many entomologists throughout its range in Britain and Ireland.

Stethophyma grossum (Linnaeus, 1758)   **Large marsh grasshopper**

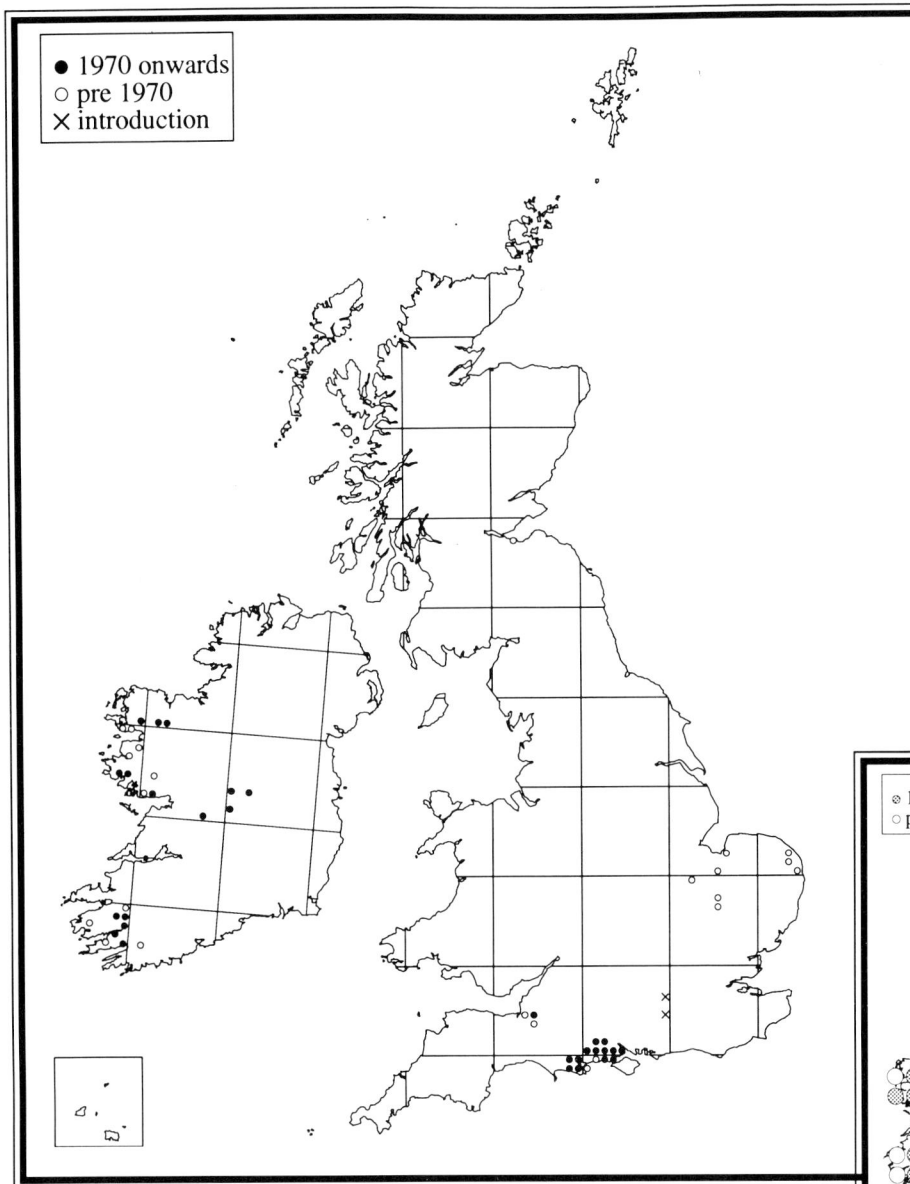

- ● 1970 onwards
- ○ pre 1970
- ✕ introduction

Marshall and Haes (1988) Description and species account pp106–107, Plate 6.

GB protection/threat status RDB2 (Vulnerable).

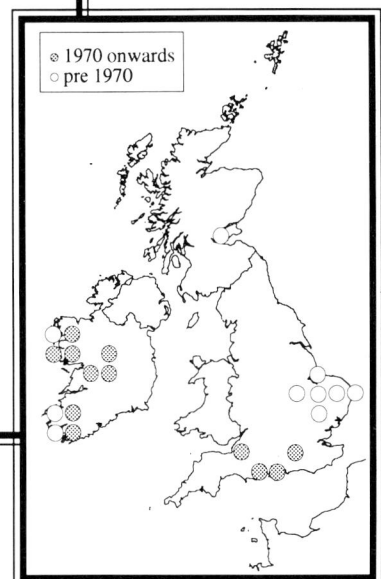

- ● 1970 onwards
- ○ pre 1970

Status

Native in southern and eastern England and in central and western Ireland, but not recorded from Wales, Scotland or the Channel Islands.

This is our largest (22–36 mm) native grasshopper; females are distinctly larger than males. Colour varies from yellowish green to olive-brown, with red underneath the femora of the hind legs and a plum-coloured form is sometimes observed in females. The 'song' is a brief, intermittent clicking which is unmistakable and can be heard at some distance. It is unlike the conventional stridulation of other grasshoppers and sounds like someone using hedge shears in bouts of a few sharp strokes. It is usually heard only in hot weather. Adults fly strongly when disturbed. Nymphs emerge in late spring and adults appear from mid-July onwards. Adults can survive into early November. This is a wetland species, usually found in acid bogs with tussocky grass, especially purple moor-grass, and scrub, such as bog-myrtle (*Myrica gale*) and cross-leaved heath. It occurs at undisturbed sites in England, such as quaking bogs in the New Forest, the Dorset heaths, and (now only as a rarity) on the Somerset Levels 'heaths and moors', many of which are protected. It has been lost entirely from East Anglia where it formerly occurred in the Norfolk Broads, and in the area of Whittlesea Mere in Cambridgeshire. There are even earlier records from Thames-side, suggesting that it may have been lost from other riverside wetlands. Its occurrence in western Ireland is less localised, reflecting the survival of many more suitable wetland sites than in England; these include some nationally important bogs as well as lakeside and riverside sites (see for example Foss & Speight 1989).

Oedipoda caerulescens (Linnaeus, 1758)

Blue-winged grasshopper

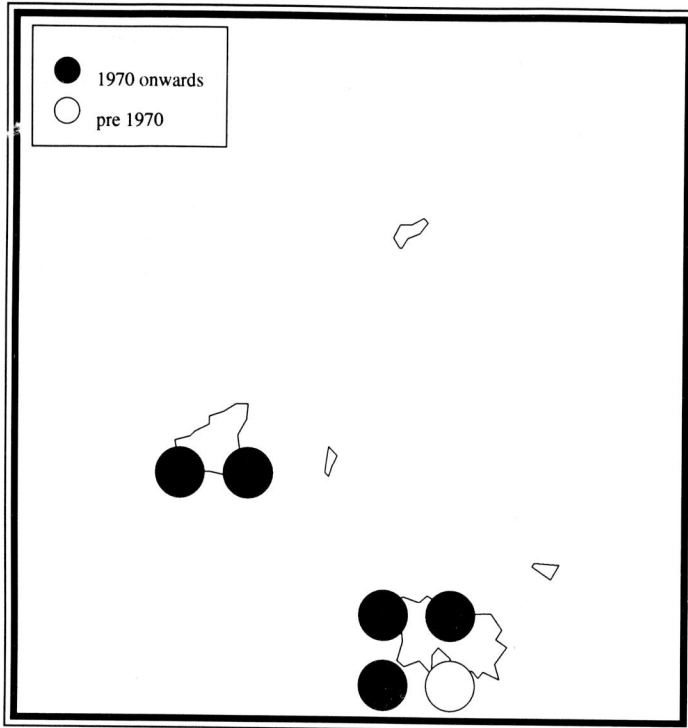

- ● 1970 onwards
- ○ pre 1970

Marshall and Haes (1988) Description and species account p105, Plate 6.

Status

Native on the Channel Islands, and recorded once (probably as a stray) on the Isles of Scilly. In continental Europe, it occurs from the Mediterranean to southern Sweden, and also in north Africa and Asia.

This is a cryptically and variably coloured, rather large (17–26 mm) grasshopper. In flight it is very distinctive, when bright blue basal patches with dark borders on the hindwings are shown. Flight is often prolonged, but on landing the grasshopper instantly merges into the background. Stridulation is usually only heard during courtship – a soft, short buzzing. It occurs mainly on coastal sand dunes, south-facing cliffs, disused quarries and stony fields. Adults appear in early July and can be found through to October and early November.

Euchorthippus pulvinatus (Fischer de Waldheim, 1825) subspecies *elegantulus* (Zeuner, 1940) *

Jersey grasshopper

Status This subspecies apparently occurs only on the south coast of Brittany and in the south-western part of Jersey. Elsewhere in western Europe the subsp. *gallicus* Maran occurs and the nominate subsp. *pulvinatus* is found in eastern Europe and Asia. The Jersey population, from which subsp. *elegantulus* was described by Zeuner, is isolated from all others.

A slender grasshopper (10–22 mm) which is usually brownish or straw-coloured. Both sexes are winged and fly readily, although the wings are short. Nymphs hatch in May and adults are present from late June to October. Stridulation is a series of brief 'zzip' sounds at the rate of one per second for up to 30 seconds. It has been found in hot sunny situations near the sea, on dune systems, roadside verges and unimproved pastures, often where fully exposed to Atlantic gales. Some sites appear to be vulnerable to increased recreational pressures.

* Note the correct attribution of authorship given here; that in Marshall and Haes (p70) gives an incorrect date.

Marshall and Haes (1988) Description and species account p117, Plate 7.

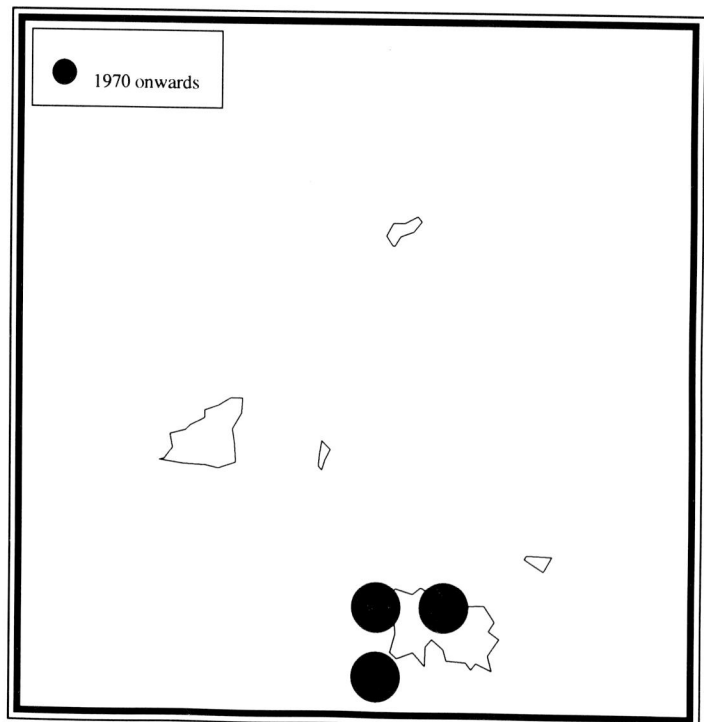

- ● 1970 onwards

Stenobothrus stigmaticus (Rambur, 1839) **Lesser mottled grasshopper**

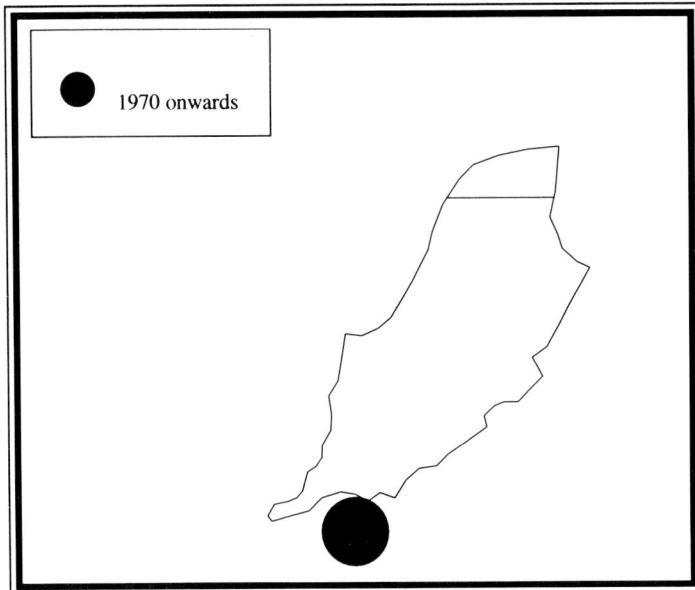

1970 onwards

Marshall and Haes (1988) Description and species account pp108–109, Plate 6.

IoM protection/ threat status Wildlife Act 1990 Schedule 5.

Status

Isle of Man only. Known to occur in northern France, Belgium, The Netherlands, Germany and the Baltic coast.

This, our smallest grasshopper (10–15 mm), is usually green with brownish wings, sometimes developing orange-red coloration on the tip of the abdomen. Eggs usually hatch in May and adults appear in early July, going through to mid-October. Stridulation is similar to that of *Chorthippus parallelus* – an unobtrusive series of chirps in bursts of 1–4 seconds' duration. It occurs on the Langness peninsula at the southern end of the Isle of Man, on dry, grassy slopes above the sea, around rocky outcrops and in a small area of vegetated sand dunes. Its preferred vegetation is short turf (usually 5–10 cm tall), often with heather (*Calluna vulgaris*) or western gorse (*Ulex galii*). This short, grazed turf appears essential to its survival, and lack of grazing in recent years may threaten its survival on the Isle of Man (Cherrill 1994).   It has been recorded at Langness regularly since it was first discovered there in 1962 (Ragge 1963) and, despite being a totally isolated population, there are strongly held opinions that the species is native to the Isle of Man and is not an accidental introduction (Ragge 1965; Burton 1990).

Stenobothrus lineatus (Panzer, 1796)   **Stripe-winged grasshopper**

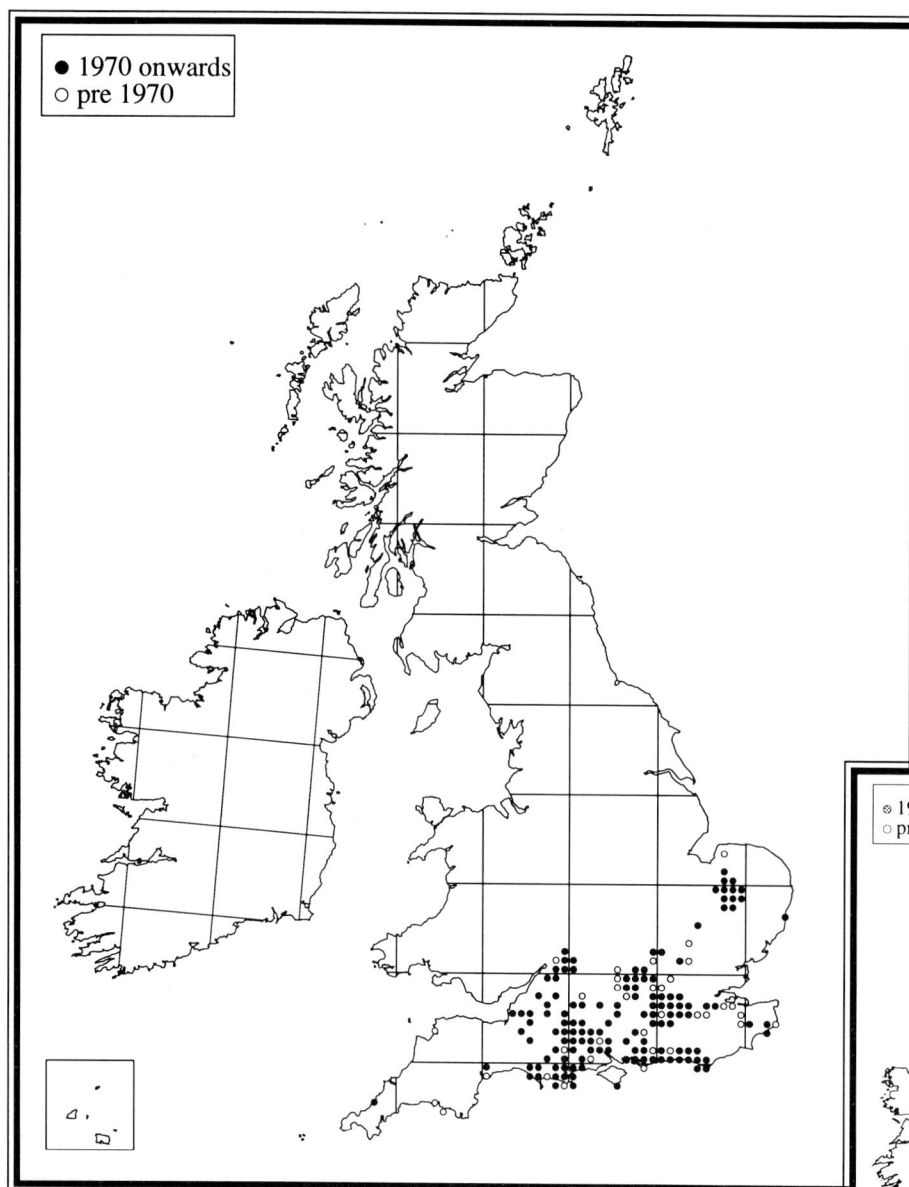

Marshall and Haes (1988) Description and species account pp107–108, Plate 6.

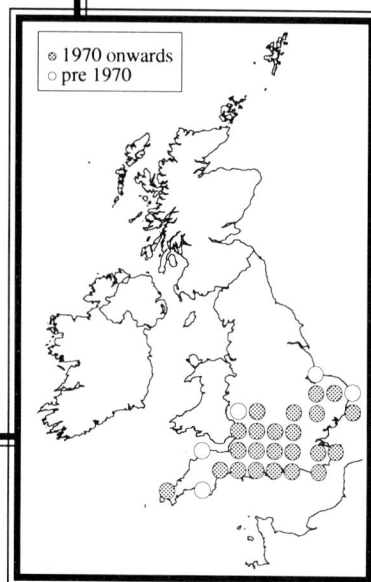

Status

Native in southern and eastern England. Not recorded from Wales, Scotland, Ireland or the Channel Islands.

This medium-sized (15–23 mm) grasshopper is quite brightly coloured, usually green and brown, with red on the abdomen, but purple or pinkish colour forms can occur. A distinctive feature of this species is the white line and elongated spot (*stigma*) on the forewing. Stridulation is a short, pulsating, metallic rasp; the distinctive courtship song is more sustained. Nymphs appear in May and adults are present from late June onwards, surviving until October. It is a species of short turf, often with bare ground, mainly on chalk and limestone, but also on sandy soils on dry heaths. It is characteristic of long-established and stable, species-rich grasslands, a habitat which it shares with several species of 'blue' butterfly (Lycaenidae). Many of the best sites for *S. lineatus* are protected for their botanical interest or for butterflies. It appears to be subject to natural fluctuations in numbers: in hot summers local populations can become large, only to decline again and become difficult to find after a series of poor summers. It occurs at suitable sites from East Anglia to Cornwall: on grasslands on chalk and Jurassic and Carboniferous limestones, on heaths in Breckland, Surrey, Hampshire and Dorset, and on some sand dune systems. The largest populations are on the North and South Downs. It is uncertain whether or not this species has increased its range in recent decades: more localities are now known than before 1960, but this is probably a result of increased activity by recorders since about 1975.

Omocestus rufipes (Zetterstedt, 1821)   **Woodland grasshopper**

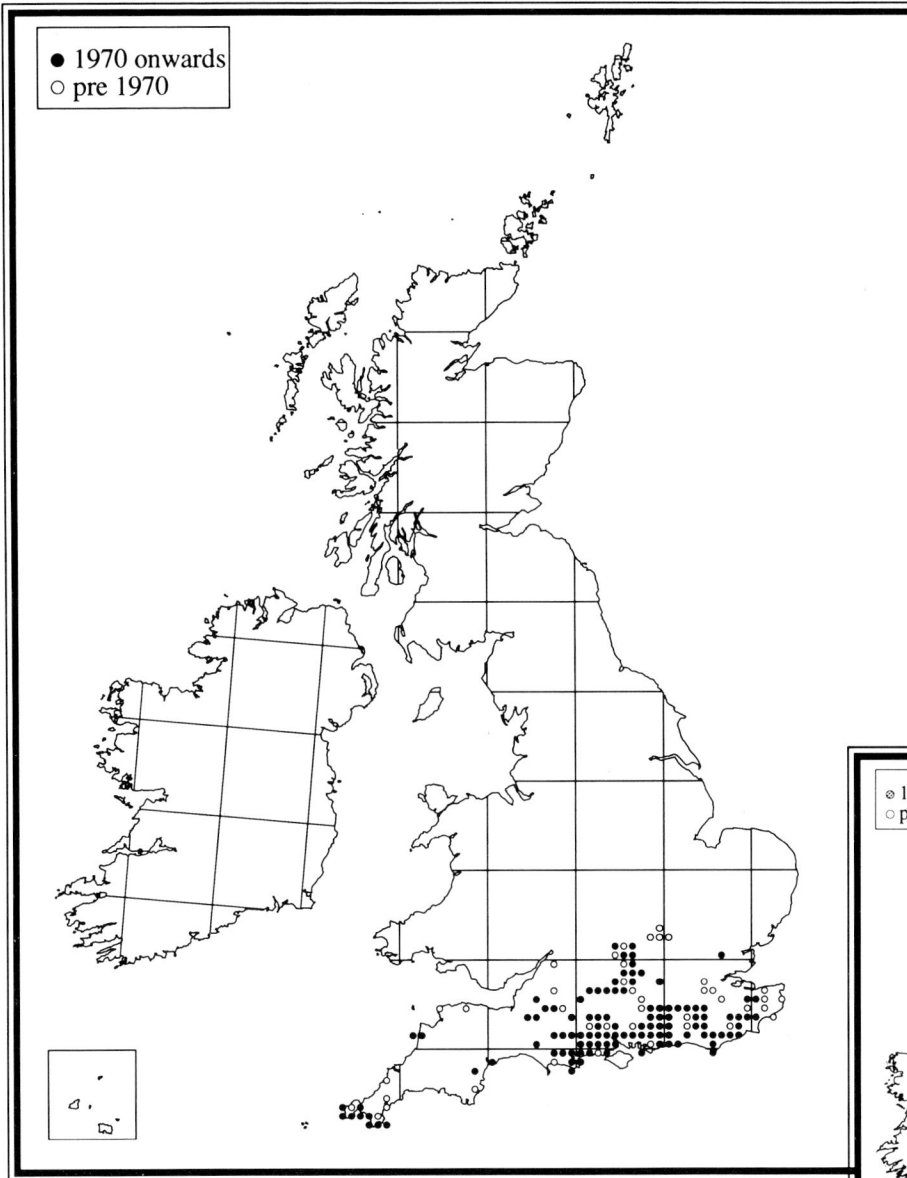

Legend:
- ● 1970 onwards
- ○ pre 1970

Marshall and Haes (1988) Description and species account pp109–110, Plate 6.

GB protection/threat status Nationally Scarce (B).

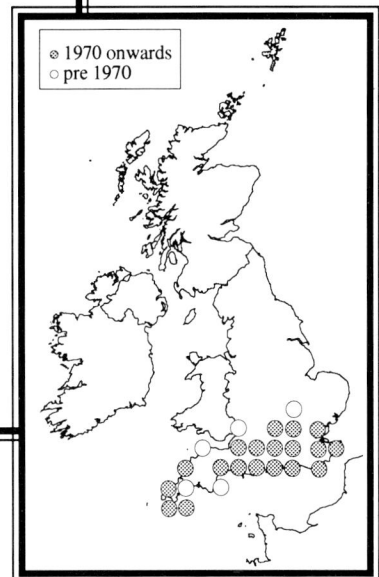

Inset map legend:
- ⊘ 1970 onwards
- ○ pre 1970

Status

Native in southern England. This species has been misidentified in the past and there are pre-1960 records from Suffolk, Cambridgeshire, N England, Wales and even Scotland, which either have been disproved (see Kevan 1956; Pickard 1956) or must be regarded as uncertain. Not recorded from Ireland, the Isle of Man or the Channel Islands.

This is a medium-sized (12–20 mm), dark, greyish brown grasshopper. The underside of the posterior of the abdomen in mature adults is red or orange. Females exhibit two colour variations, which are either all brown, or with green forewings and upper side of the head and pronotum. Its most distinctive feature is the palps, which have chalk-white tips. Colour forms of other species, especially *Chorthippus brunneus* and *Omocestus viridulus*, have been confused with this species. Nymphs emerge in April or May and become adult during June; adults rarely survive beyond mid-October. Stridulation is a sibilant clicking, starting quietly and increasing in volume before ending abruptly, lasting 5–10 seconds: it is similar to, but shorter than, that of *O. viridulus*. It is a grass-feeding species which occurs in rides and clearings in woodland and on grassland or heath near woodland or scrub, persisting even when scrub encroaches if there is still some grass present, and the site does not become too shaded. In Cornwall it occurs on heathy areas, including sea cliffs, well away from any remaining woodland. It is most common in the New Forest and on the western side of the Weald and in the extensive woodlands and plantations in East Kent. There are few records from north of the Thames, possible records from the Cotswolds, doubtful records from the Forest of Dean and Wales, and no records from the Isle of Wight. The few isolated, and usually small, populations north of the Thames and in the West Country have high local importance for nature conservation.

Omocestus viridulus (Linnaeus, 1758)　　Common green grasshopper

Marshall and Haes (1988) Description and species account pp110–111, Plate 6.

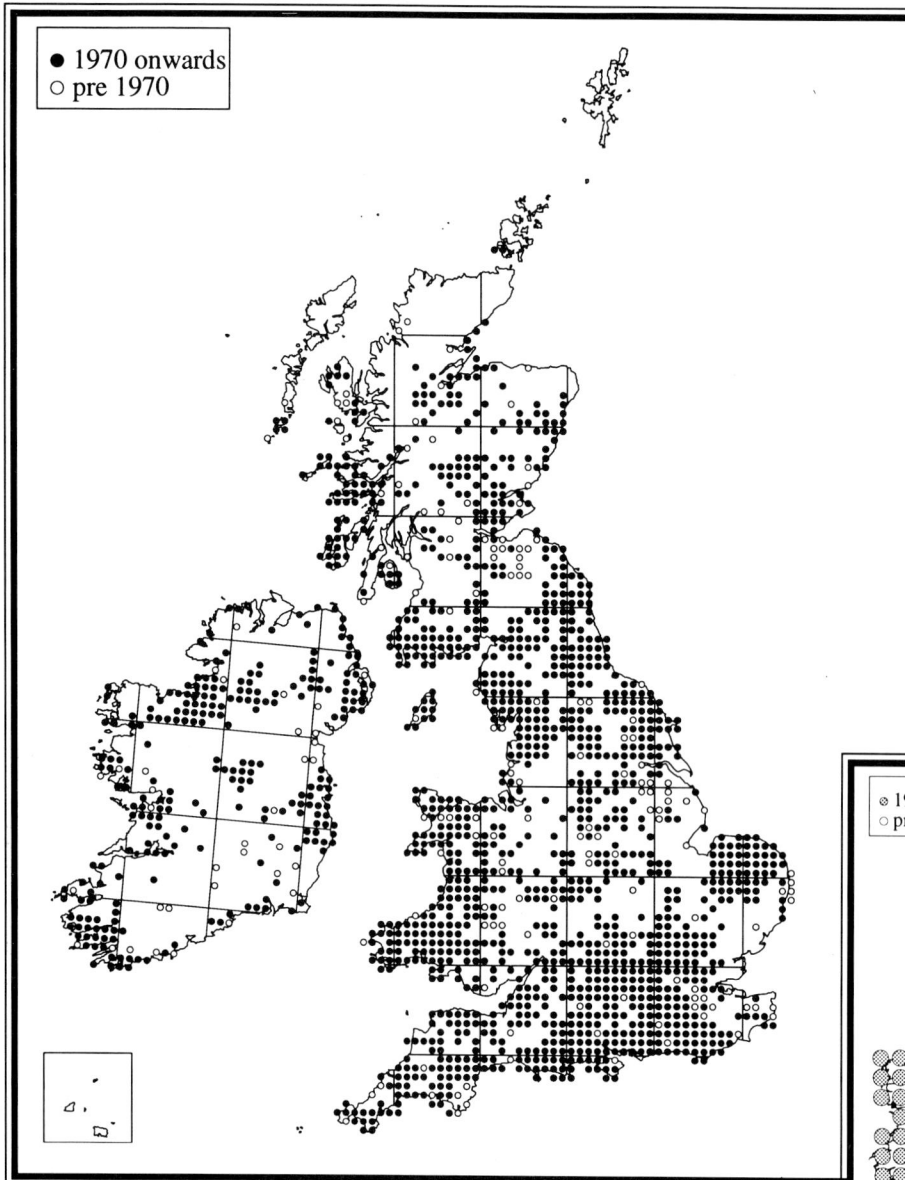

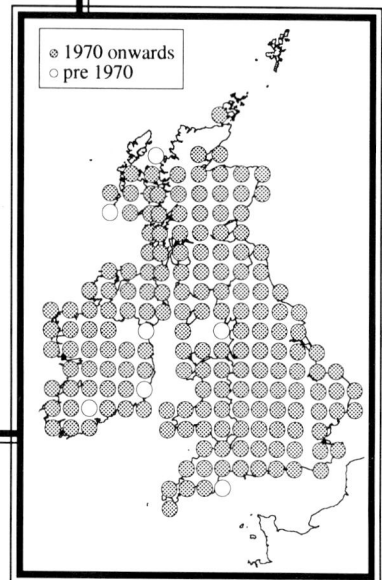

● 1970 onwards
○ pre 1970

⊘ 1970 onwards
○ pre 1970

Status

Despite being our most widespread orthopteran, it is absent from the Channel Islands, the Isles of Scilly and Shetland, and apparently from parts of the mainland of both Britain and Ireland. Some absences from the respective mainlands may have more to do with coverage of recording than with the rarity or absence of the species, especially in Ireland where it is the commonest grasshopper in areas where surveys have been made.

A medium-sized (15–22 mm) grasshopper, usually green or green and brown, without red or orange colour on the abdomen. The palps may be pale, but never chalk-white. Nymphs hatch in April and May and develop rapidly to adults by mid-June (even late May in some recent hot summers), which sometimes survive into November. In hot weather, adult males fly readily. Stridulation is a rapid, prolonged, sibilant clicking which starts quietly and rapidly increases in volume, and is sustained for 10–20 seconds or even longer. It is one of the familiar sounds of summer. Even longer songs are produced in courtship. It favours quite long grass, often in moist situations, and particularly old, unimproved grasslands which are not heavily grazed or mown, including parkland. Where the soil is normally damp, it appears to be able to tolerate shorter vegetation. However, it does not seem to be as capable as *Chorthippus parallelus*, with which it often occurs, at colonising new roadside verges and other waste, grassy places. It is also conspicuously absent from some coastal locations, particularly the low rainfall areas of east and south-east England with free-draining soils, but also some of the Devon and Cornwall coastline. However, it does occur on many west coast islands. It is the most numerous grasshopper in upland areas, and the only species regularly found above 500 m.

Chorthippus brunneus (Thunberg, 1815)   **Field grasshopper**

Marshall and Haes (1988) Description and species account pp112–113, Plate 7.

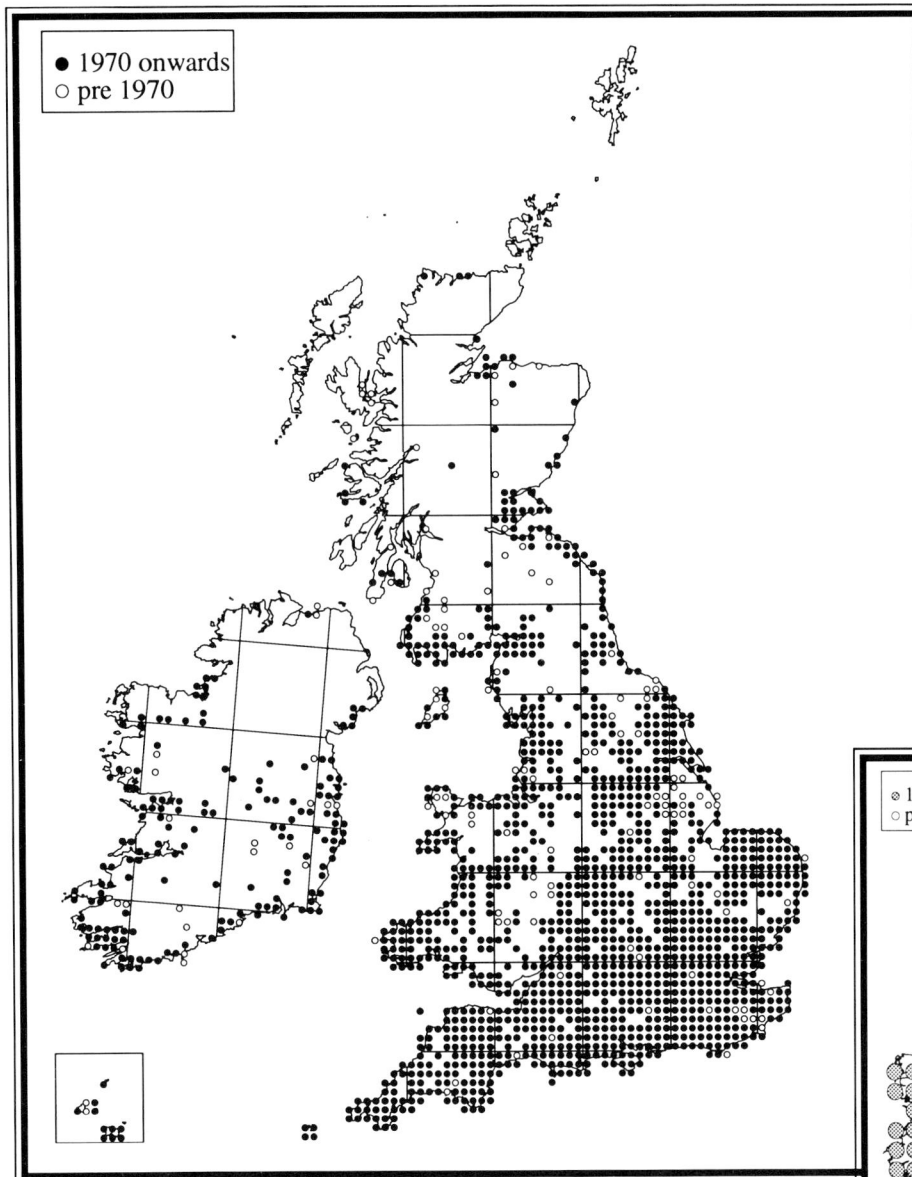

Legend: ● 1970 onwards ○ pre 1970

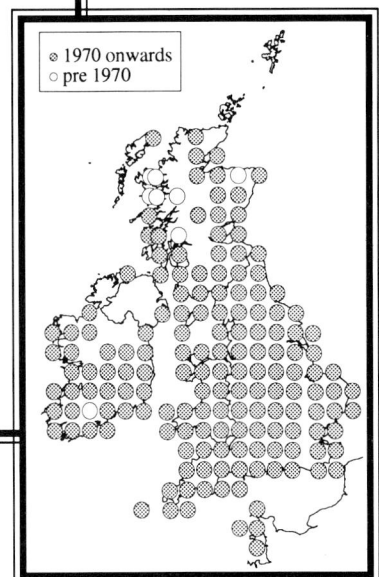

Legend: ● 1970 onwards ○ pre 1970

Status

Widespread in southern Britain and Ireland, but restricted to coastal areas in the north of its range. It has colonised many southern, often small, offshore islands, including all the major Channel Islands, the Isles of Scilly, Lundy, Clear Island and Sherkin Island. This distribution pattern is known for other organisms at the north-western edge of their European range.

The colour of this medium-sized (15–25 mm) basically brownish grasshopper varies considerably, with striped and mottled forms and with parts of the body varying from buff through orange to purple. The underside of the thorax is distinctively hairy in both sexes. It has a very long season with the hatching period beginning in late March in warm springs and with adults surviving to early December in favourable locations after a mild autumn. Both sexes are capable of active flight. Stridulation is a short, brisk chirp repeated at short intervals. Two or more males will often respond to each other, alternating rapidly. Again, this is a familiar sound of summer, even in urban areas. The preferred habitat is short vegetation in dry, sunny situations where there is no shade; sites such as downland and coastal grassland are typical for this species. Possibly because of its ability to fly well, it is often found on quite isolated patches of suitable vegetation, for example on roadside verges and on waste ground and railway cuttings and embankments, and even in parks in built-up areas. It is usually absent from very damp sites such as heathlands, wetlands and mountains.

Chorthippus vagans (Eversmann, 1848)

Heath grasshopper

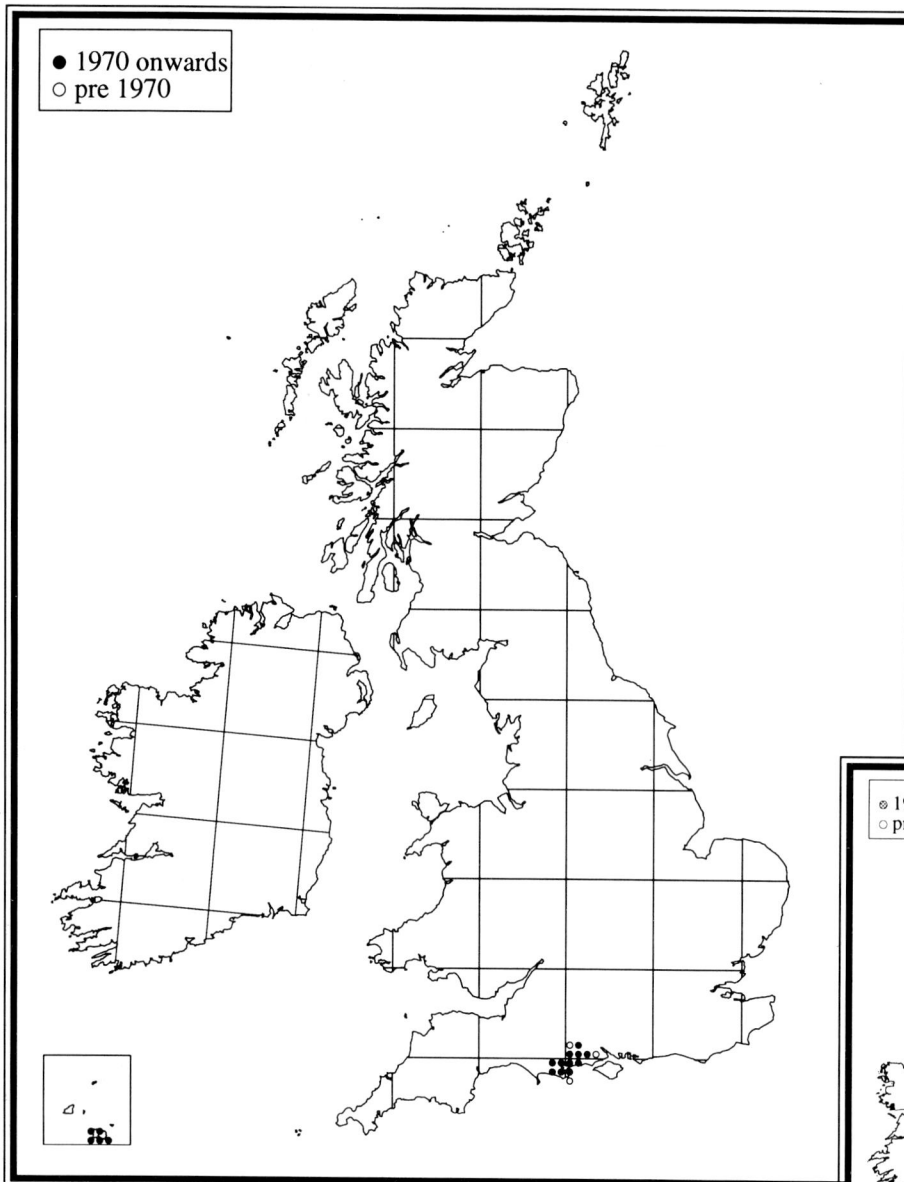

Marshall and Haes (1988) Description and species account pp113–114, Plate 7.

GB protection/threat status RDB3 Rare.

● 1970 onwards
○ pre 1970

⊗ 1970 onwards
○ pre 1970

Status

Restricted to a small area of southern England, on heaths in Dorset and Hampshire, and Jersey in the Channel Islands. First identified from British material in 1921, and the first definite, localised record was from near Studland, Dorset, in 1933. Possibly the commonest grasshopper on Jersey where, as in continental Europe, it occurs in a range of habitats.

This medium-sized (13–21 mm) species is superficially similar in appearance to colour forms of several of the more common grasshoppers, so that identification must be made with care. In particular, female *Mymeleotettix maculatus* can be confused with male *C. vagans* and the two species often occur together. It is dark, greyish brown, with little variation. In mature adults the abdomen develops an orange/orange-red tinge; this colour extends to the hind femora, which also have two dark bands. It has a short season, with nymphs hatching in May and adults usually appearing in late June and July, but surviving only to early October. Stridulation is similar to that of *C. parallelus*, but slower and louder, lasting about 5 seconds. It is found only in dry heathland, frequently in areas of pure heathers without grasses because it feeds readily on heathers. Although a few localities are known to have been lost (eg Cranborne Common and Holton Heath, Dorset) in recent decades, it appears to be widespread on suitable heaths in East Dorset, around Poole Harbour and in the west and south of the New Forest, and some sites are protected. However, the core of its previous range must have been lost before, or soon after, its presence in Britain was recognised, mainly as a result of the expansion of the Bournemouth/Poole conurbation and the conversion of many heaths to plantations and arable farmland.

Chorthippus parallelus (Zetterstedt, 1821)   **Meadow grasshopper**

Marshall and Haes (1988) Description and species account pp114–115, Plate 7.

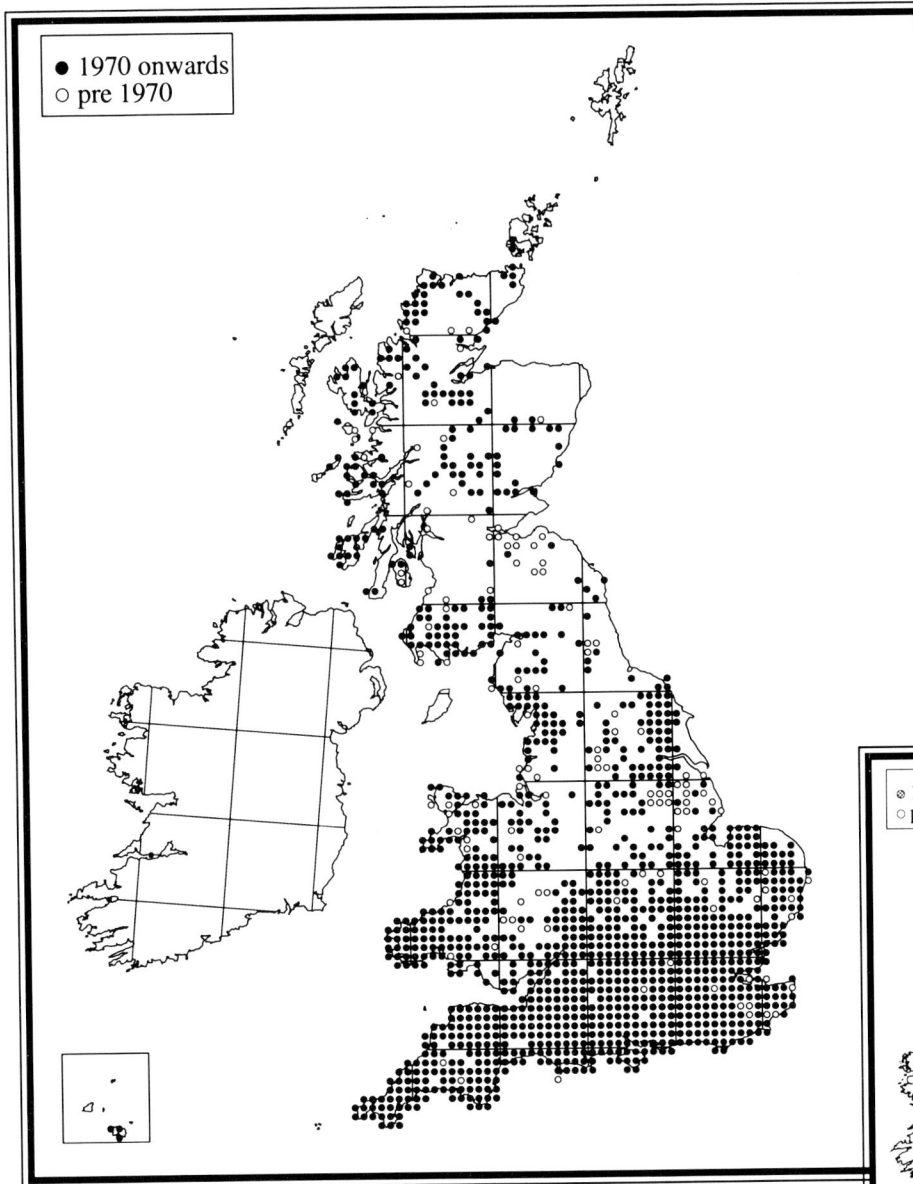

- ● 1970 onwards
- ○ pre 1970

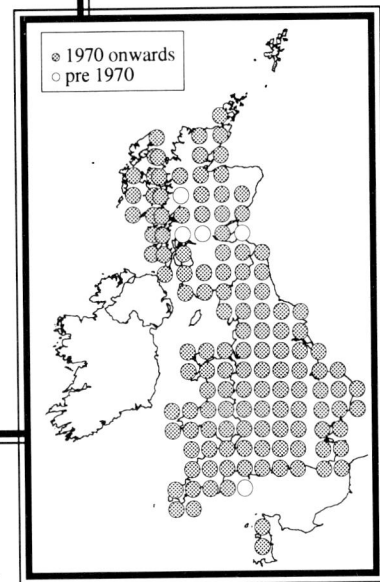

- ◉ 1970 onwards
- ○ pre 1970

Status

Distributed throughout Britain, except the Isles of Scilly, Outer Hebrides and Shetland, but in the north it seems to be more widespread in the west. Recorded at altitudes of over 800 m in the Scottish Highlands. In the Channel Islands, it is known only from Jersey. Absent from parts of the east coast, such as eastern Norfolk, the Spurn peninsula and Holderness, Yorkshire, and almost all the east coast north of Whitby; also absent from the Isle of Man and Ireland.

This medium-sized (10–22 mm) grasshopper is usually green-coloured, but often with brown wings and occasionally entirely brown-coloured, with some adult females a vivid pinkish purple. Although normally with vestigial hindwings and short forewings (females normally have very reduced forewings), a fully winged form (f. *explicatus* (Sélys)), which is capable of flight, occurs in populations usually during the latter part of hot summers. This species has a long season, with hatching beginning in late April, adults appearing in June and remaining abundant through to September,

and sometimes surviving into November. Stridulation is a brief (3–5 second) buzzing with a rapid irregular pulse, repeated at varying intervals. It is typically associated with coarse grasses in a wide range of habitats, such as sand dunes, saltmarshes, woodland rides, roadside verges, waste ground, valley wetlands and wet grassy moorland, where it may occur in large numbers. It also occurs, but in restricted numbers, in short turf, for example on heaths, cliffs and chalk downland: in such sites it is usually outnumbered by *C. brunneus* and/or *Myrmeleotettix maculatus*.

Chorthippus albomarginatus. (De Geer, 1773) **Lesser marsh grasshopper**

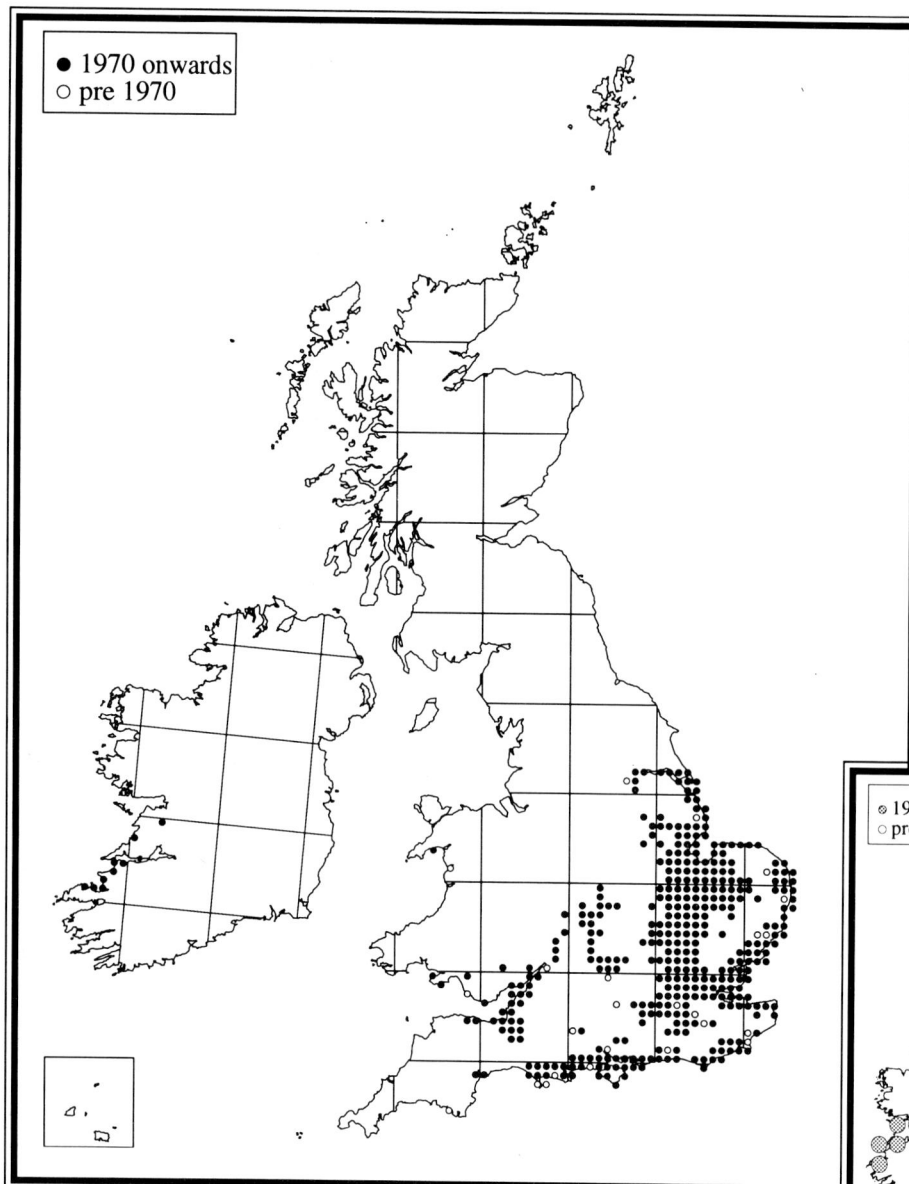

Marshall and Haes (1988) Description and species account pp115-117, Plate 7.

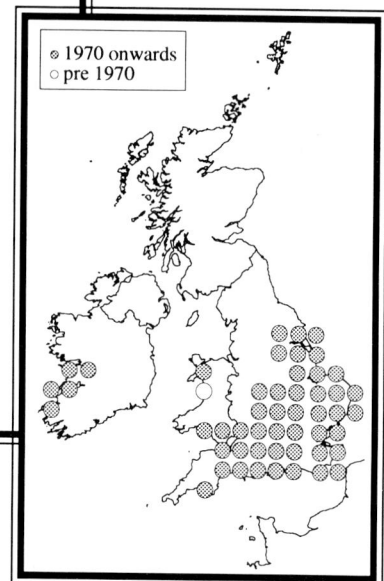

- ● 1970 onwards
- ○ pre 1970

- ◉ 1970 onwards
- ○ pre 1970

Status

There appears to have been considerable expansion of range over the last 25–30 years, particularly in inland areas along major river systems and trunk roads. Now known south of a line from southern Yorkshire to south Wales and also in south-western Ireland. Old records from north Wales, Merseyside, Morecambe Bay and Cornwall have not been substantiated, despite searches since 1960, and must be doubtful. Not recorded from the Channel Islands, Isles of Scilly, Isle of Man or Scotland.

This species is similar in appearance to *C. parallelus*: it is of medium size (14–21 mm), light green and/or straw-brown, with several colour variants, but it is less brightly coloured than *C. parallelus*. Both sexes have functional wings. It is a mid-season species, with nymphs hatching in May and adults appearing by late June; although there are a few November records, most adults die by the end of October. Normal stridulation is a gentle burr repeated in cycles 2–6 times. The courtship stridulation is a complex repeated sequence, somewhat like a watch being wound. On the coast this species is normally found on the landward side of dunes, saltmarshes and shingle banks, often in low-lying pastures and the grassy slopes of dykes (embankments). Inland it has been found in a wide range of rough grassy areas. Beside rivers these include flood meadows and wetlands, often with sedges, and also grassy embankments. It also occurs on roadside verges, waste ground, urban parkland, set-aside arable land, damp clearings in woods and even among gorse in Oxfordshire. In Ireland it has been recorded beside Lough Bunny on the eastern edge of the Burren, Clare, and from dune systems and a coastal golf course in Counties Kerry and Clare.

38

Gompboceripppus rufus (Linnaeus, 1758)

Rufous grasshopper

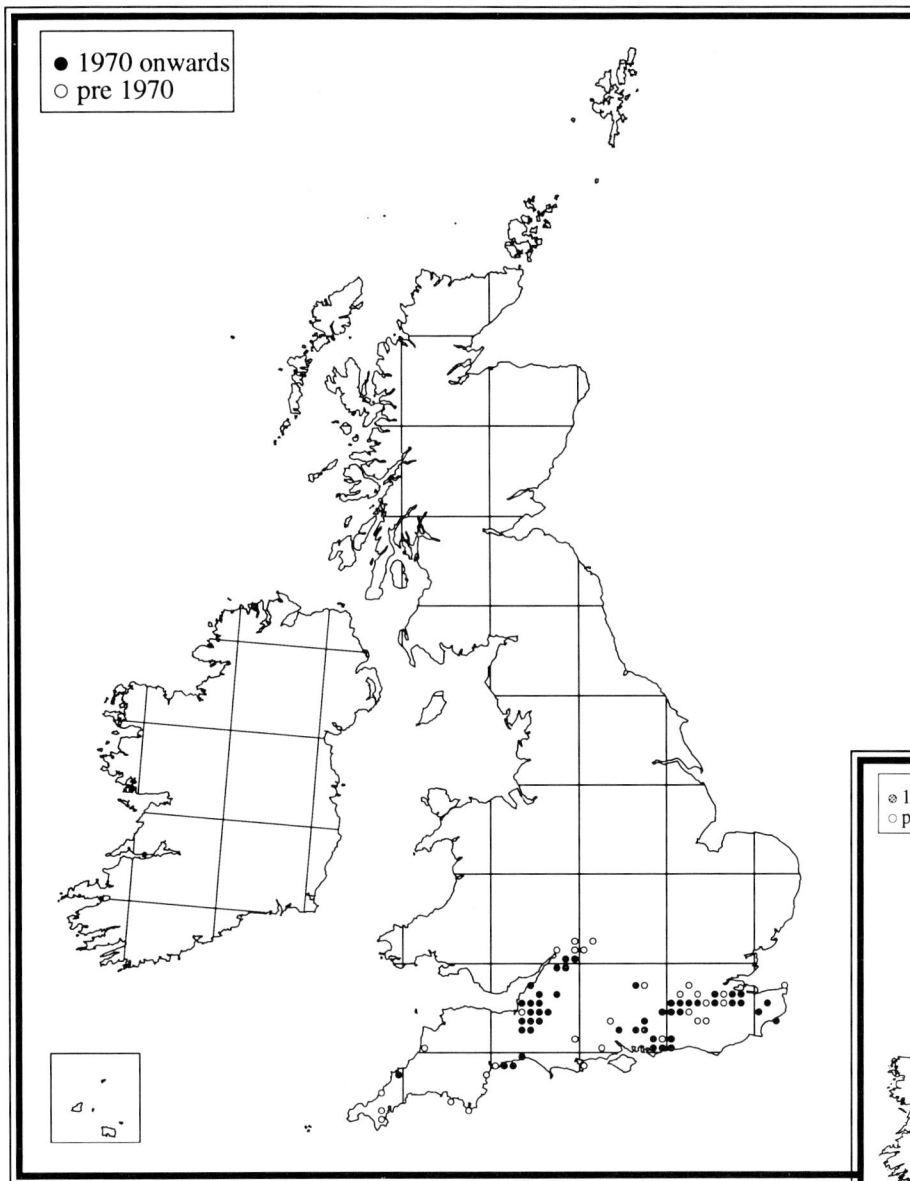

Marshall and Haes (1988) Description and species account pp118–119, Plate 7.

GB protection/ threat status
Nationally Scarce (B).

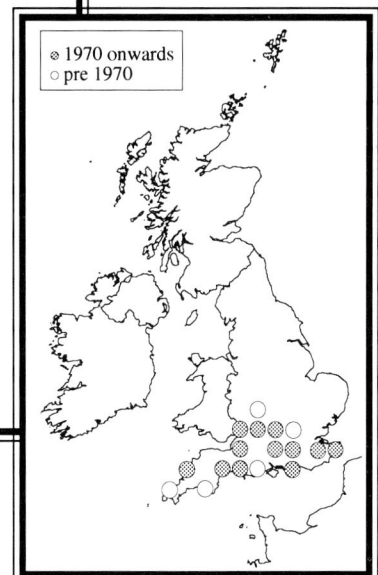

● 1970 onwards
○ pre 1970

⊗ 1970 onwards
○ pre 1970

Status

A native species at the edge of its range in southern England. One unconfirmed record from south Wales, but not recorded from the Channel Islands, Isles of Scilly, Isle of Man, Scotland or Ireland. Most records from the south-west are of isolated populations; some have not been seen for many years.

A medium-sized (14–22 mm), broad-bodied, brown grasshopper with pale/white-tipped, clubbed antennae. Mature males have an orange-red tip to the abdomen, but this coloration is less pronounced in females. Green forms do not occur, but some females are reddish purple. Nymphs hatch in late May and adults appear from late July onwards. In mild autumns, adults can survive through to early December. Stridulation is a soft but urgent buzzing, which fluctuates slightly, dying away at the end. The courtship song and accompanying display are distinctive. This species is associated with rough, dry grassland on calcareous soils (mainly on Jurassic limestones, chalk and calcareous sandstones, more rarely

on Carboniferous limestone). It is found typically on south-facing slopes and in sheltered valleys and coombes, usually where some scrub is present, or along the open margins of woods. It has also been recorded occasionally from wooded undercliffs and sand dunes. Numbers seem to fluctuate at the few sites which have been surveyed regularly – increasing in hot summers and declining in cool summers. The outbreak of the *myxomatosis* virus in rabbits *(Oryctolagus cuniculus)* in the 1950s, and the resulting decline in their numbers, probably benefited this species by allowing long coarse grassland to develop where previously there was intensively grazed, short turf.

Myrmeleotettix maculatus (Thunberg, 1815) **Mottled grasshopper**

Marshall and Haes (1988) Description and species account pp119–120, Plate 6.

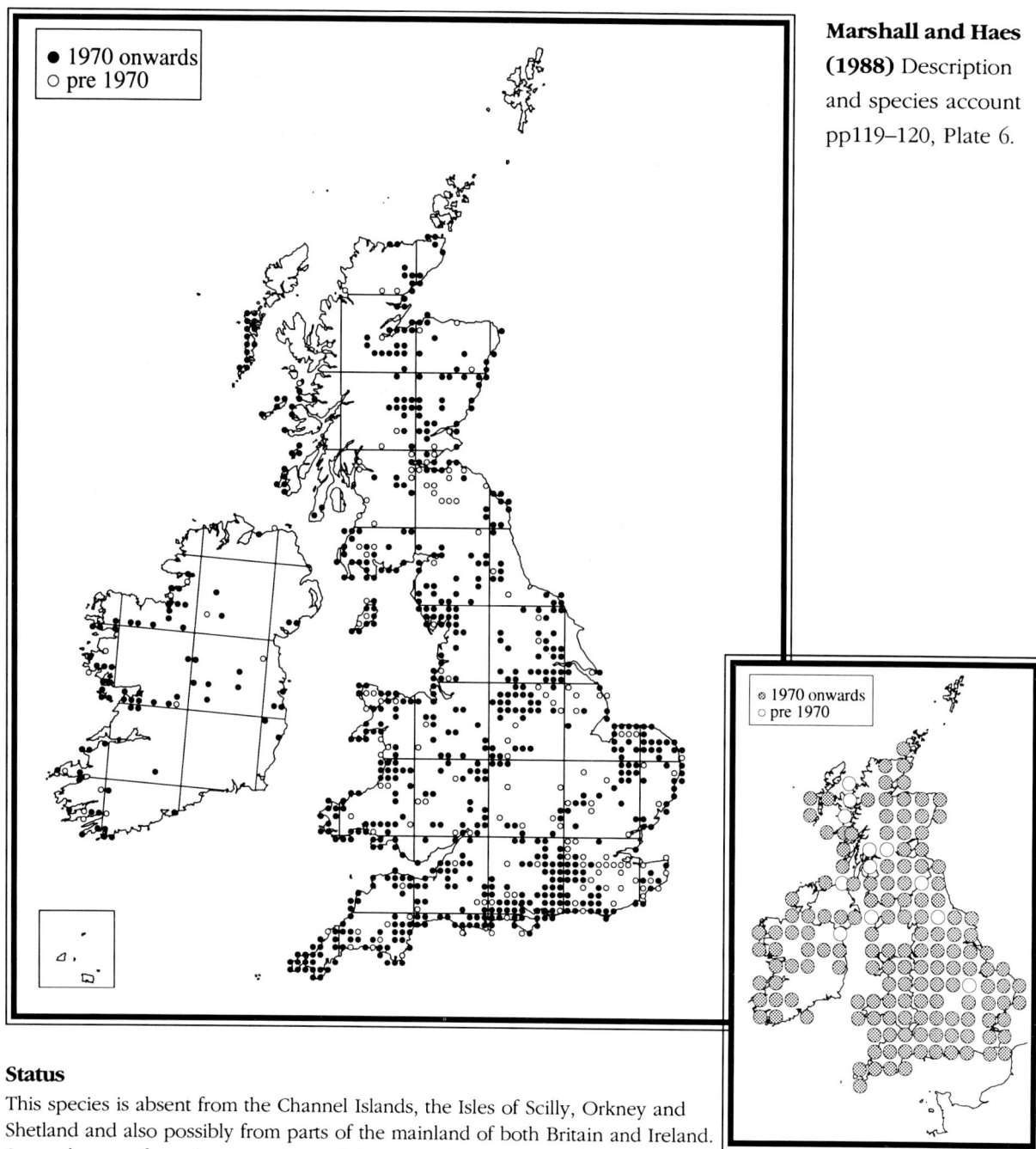

● 1970 onwards
○ pre 1970

⊗ 1970 onwards
○ pre 1970

Status

This species is absent from the Channel Islands, the Isles of Scilly, Orkney and Shetland and also possibly from parts of the mainland of both Britain and Ireland. Some absences from the respective mainlands may have more to do with coverage of recording than with the rarity or absence of the species; its apparent scarcity in Ireland is probably misleading, but as yet no records have been traced from the south-east.

A small (12-19 mm) grasshopper with 12 described colour forms. The clubbed antennae of males and thickened antennal tips of females are distinctive and both lack the white tips found in *Gomphocerippus rufus*. Nymphs emerge from the end of April to June and develop rapidly to adults by mid-June. Adults rarely survive beyond the end of October. It is a gregarious species and is capable of strong flight over short distances. Stridulation is a sequence of short buzzing sounds lasting 10–15 seconds: it begins softly, reaches a climax and then stops abruptly. As with *G. rufus*, the males have an elaborate courtship song and display. It favours a distinctive habitat – dry situations exposed to the sun with short turf and bare ground, usually on free-draining soils on sand, gravels, chalk and other limestones. Typical locations are the steep sides of disused quarries, road and rail cuttings, heaths and coastal dunes. In the north and west of its range (including Ireland), it has been recorded in tiny pockets of suitable habitat surrounded by seemingly unfavourable moorland and rough grassland, but more extensively on machair on western coasts. It is present on several small, off-shore islands.

Dictyoptera – cockroaches and praying mantids

This large order consists of two distinctly different sub-orders: the Blattodea (cockroaches) which typically are dorso-ventrally flattened, with a beetle-like appearance; and the Mantodea (praying mantids) which are like very elongated grasshoppers with greatly enlarged and modified forelegs that are used in catching prey.

Blattodea – Cockroaches

With over 4000 species in the world, cockroaches are an important group, but only three species are truly native to Britain and none are native to Ireland (see pp?–?). Many cockroach species are native to continental Europe, but only a few have ever been recorded as casual introductions to Britain or Ireland, as nymphs, adults or egg cases (oothecae), mainly with imported fruit and vegetables from the Mediterranean region.

There are, however, several species that are now regarded as being cosmopolitan and some have been recorded regularly in Britain and Ireland. These species originated mainly in tropical, sub-tropical or even warm temperate regions and have been spread accidentally by man to many parts of the world. They can occur almost anywhere in conditions where there is artificial heating, and some, especially *Blatta orientalis*, have been recorded out of doors, mainly in domestic and industrial refuse tips.

Some species have become firmly established as breeding populations in large, permanently heated buildings such as hospitals, hotels, restaurants, large blocks of offices and flats, and also in factories, warehouses, glasshouses, sewers and underground railway systems. Several species are so successful that they have become pests, causing damage to the fabric of buildings, carrying the risk of disease and with a strong nuisance value. Modern pest control measures are only partly successful – some species of cockroach appear to be remarkably resistant to a wide range of insecticides and even to the effects of atomic radiation.

Marshall and Haes (1988) described and illustrated the following species which are the most commonly encountered, non-native cockroaches in Britain and Ireland, all of which have become established and breed under artificial conditions.

Pycnoscelus surinamensis (Linnaeus, 1758) Surinam cockroach (pp123–124, Plate 8)

Blatta orientalis Linnaeus, 1758 Common or oriental cockroach (pp124–125, Plate 8)

Periplaneta americana (Linnaeus, 1758) American or ship cockroach (pp125–126, Plate 8)

P. australasiae (Fabricius, 1775) Australian cockroach (pp126–127, Plate 8)

Blatella germanica (Linnaeus, 1767) German cockroach (pp127–128, Plate 9)

Supella longipalpa (Fabricius, 1798) Brown-banded cockroach (pp128–129, Plate 9)

In addition, Marshall and Haes (1988) (pp147–148) listed more than 15 species that have been recorded at some time in Britain or Ireland. Of these, *Periplaneta brunnea* Burmeister, 1838, the brown cockroach, became temporarily established in the 1960s at London Heathrow Airport (Ragge 1973), and *Nauphoeta cinerea* (Olivier, 1789), the cinereous cockroach, has become established recently at Jersey Zoo (M Barclay, pers. comm.).

The recording scheme has not sought detailed records of the non-native species of cockroach and they are not mapped here. Generalised (vice-county) distribution maps were given by Marshall and Haes (1988) for the six commonest species listed above.

Mantodea – Praying mantids

Only one species (*Mantis religiosa* (Linnaeus, 1758)) of this exotic-looking group has been found in the wild, on a single occasion in 1959 in East Sussex, where it is thought to have arrived as a result of strong winds (Ragge 1965). This species was described briefly and illustrated by Marshall and Haes (1988). According to Binet (1954), *M. religiosa* occurs as far north as the Le Havre district of northern France, so it is perhaps surprising that there have not been more records of casual occurrences in southern Britain.

Several species of praying mantis are widely kept as pets, but most are tropical in origin and as yet there are no records of accidental escapes or deliberately released species becoming established in Britain or Ireland.

Ectobius species – Native cockroaches

The three native cockroach species are much smaller and delicate-looking than most of the robust, cosmopolitan species that may be more readily encountered in public buildings. None is more than 11 mm long, two species are light brownish-coloured, and one is much darker. Although they live in a range of natural habitats, they are easily overlooked as they avoid any form of disturbance and rapidly disappear into vegetation or leaf litter. They appear to be neglected by most entomologists and, as a consequence, the distribution shown for these species is almost certainly incomplete. All three species are found in southern England and most populations are discrete and isolated. There are no records of *Ectobius* spp. from Scotland or Ireland, despite the fact that on the continent *E. lapponicus* has been recorded as far north as Lapland and into eastern Russia. Although a thorough study of the ecology and biology of *Ectobius* species in Britain was undertaken in the late 1960s by Professor Val Brown (see references cited by Marshall and Haes (1988)), more remains to be discovered about the factors limiting the range of these species in Britain.

Distribution of records of native cockroaches

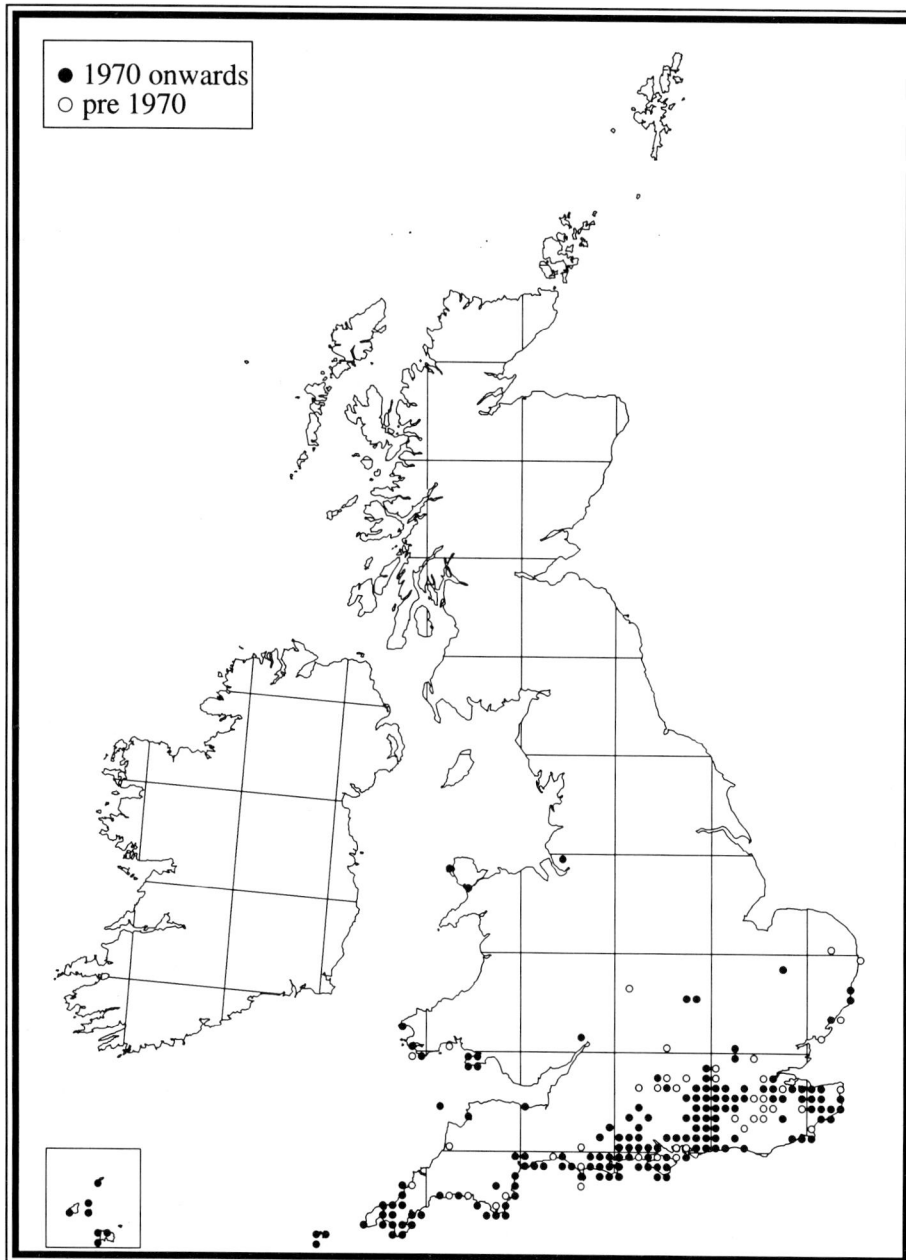

Species accounts and distribution maps

Ectobius lapponicus (Linnaeus, 1758)

Dusky cockroach

Marshall and Haes (1988) Description and species account pp129–130, Plate 9.

GB protection/threat status Nationally Scarce (B).

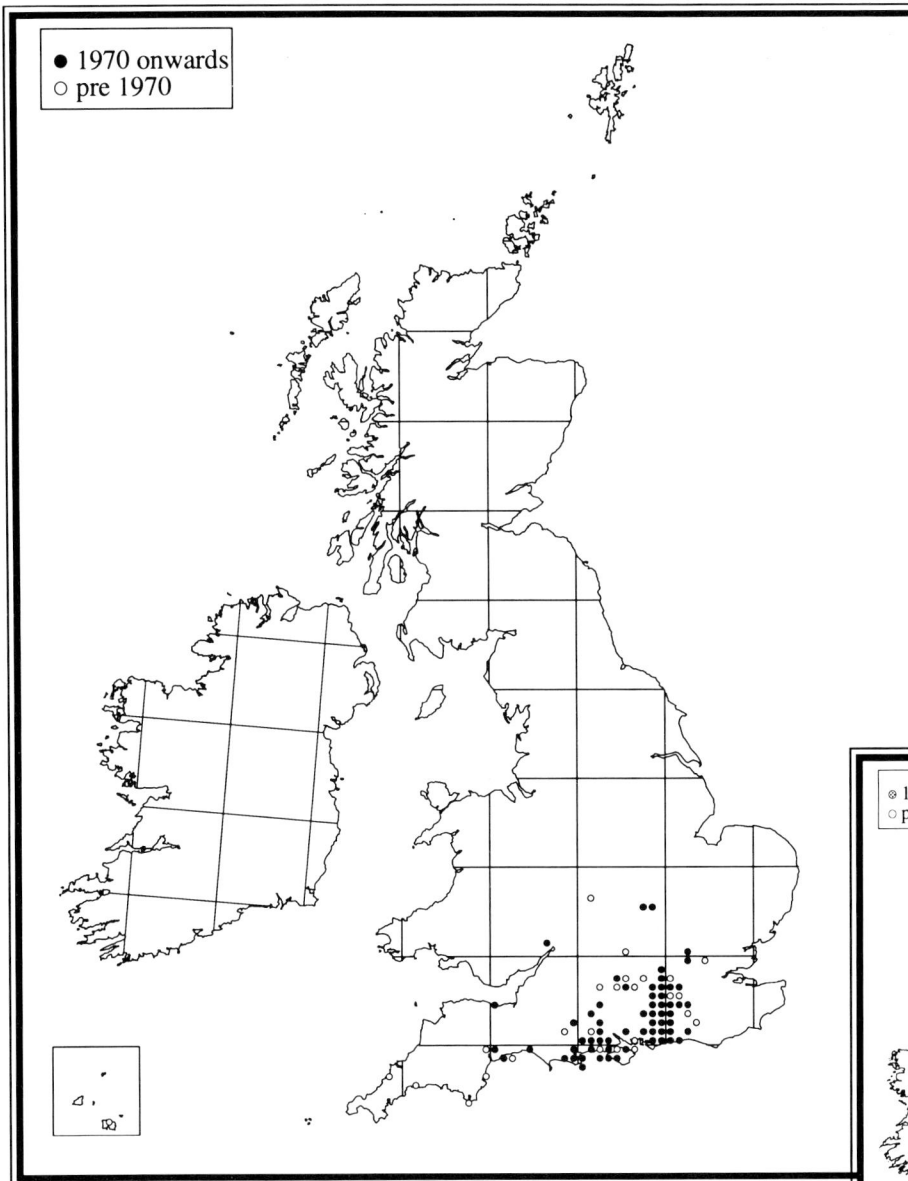

● 1970 onwards
○ pre 1970

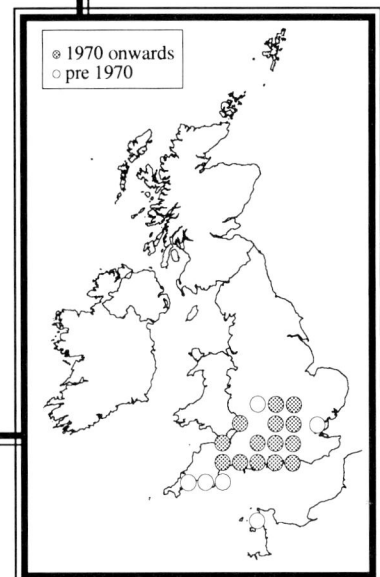

⊗ 1970 onwards
○ pre 1970

Status

Native in central southern England, with outlying records north to Warwickshire and Northamptonshire and in the west in Somerset and, before 1921, on the coast of Cornwall, but apparently absent from the extreme south-east including the whole of Kent. Recorded once from Jersey, but its status there is uncertain.

This is the largest of our native cockroaches (7–11 mm). Although the overall impression is of a light, even greyish, brown animal, both sexes have darker colouring – males on the pronotum and females on the underside of the abdomen. Males are fully winged and fly freely. Because of this, and as they are active both by day and by night, this species is the most readily encountered native cockroach, and adult males are commonly seen basking on the tops of grass stems on sunny afternoons. Females have only vestigial wings and are incapable of flight. This species has a two-year life cycle with five nymphal instars. Adults are present from the end of May to the end of September. Nymphs normally overwinter in the fourth instar and may be found in grass tussocks and deep leaf litter from October to March. Its preferred habitat is difficult to define, having been recorded from scrub and coarse vegetation on woodland margins, commons and road verges on a wide range of soil types, but it is certainly associated with scrub and grassland on both heaths and chalk. It can commonly be beaten from the lower branches of trees and shrubs, especially where grass stems grow through the branches. Its distribution appears to be centred on the New Forest and Surrey, but there are records from surrounding areas, such as the Dorset heaths, the Isle of Wight and eastern Hampshire and West Sussex. It occurs at coastal locations, especially vegetated cliffs, in south-east Devon and eastern Dorset, and at a site in South Somerset. It has also been found at a site on the edge of the Forest of Dean, Gloucestershire. North of the Thames, it is known from a few very scattered locations including Stoke Common, Buckinghamshire, Southgate Cemetery and Potters Bar, Hertfordshire, and Salcey Forest, Northamptonshire.

Ectobius pallidus (Olivier, 1789)

Tawny cockroach

Marshall and Haes (1988) Description and species account pp130–131, Plate 9.

GB protection/threat status Nationally Scarce (B).

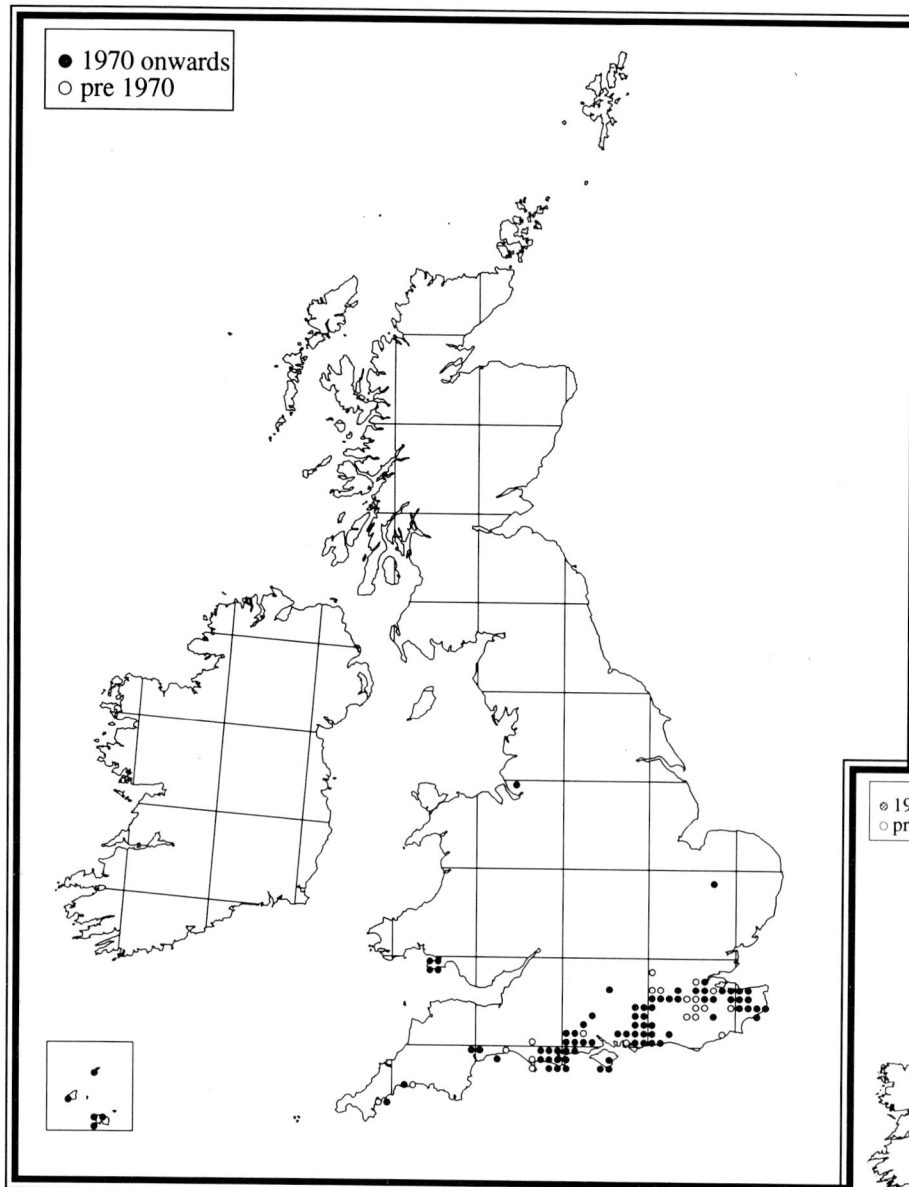

● 1970 onwards
○ pre 1970

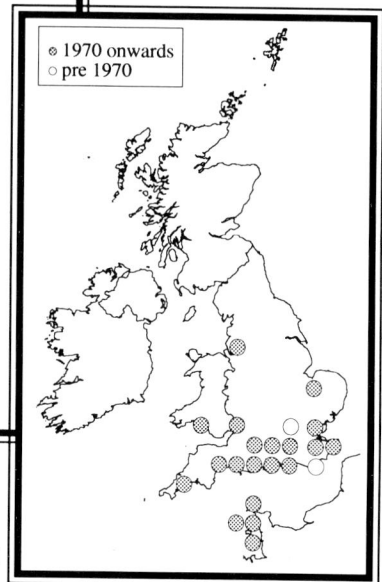

⊗ 1970 onwards
○ pre 1970

Status

Native in the extreme south and south-east of England, with a few scattered sites in the south-west, and from the Gower peninsula of south Wales and the Channel Islands. Also known from single localities in the vice-counties of West Suffolk and South Lancashire.

Of our native cockroaches, this species is medium-sized (8–9.5 mm) and is a distinctive golden brown colour. Both sexes are fully winged and are capable of flight, especially in warm sunshine, but *E. pallidus* is more active than the other two species at night and has been recorded at lepidopterists' light traps. It has a two-year life cycle, with six nymphal instars. Adults have a slightly later season than *E. lapponicus,* being present from late June until October and even into November. Nymphs normally overwinter in the fourth instar stage, resuming their development in April. It occupies a wider range of habitats than the other two *Ectobius* species, but can occur with *E. lapponicus* in scrubby woodland margins and other areas of scrub, bracken (*Pteridium aquilinum*) and rough grassland. On heaths it can be found in short heather, as well as more scrubby areas, and on downland it has been recorded from both short and long turf. In the west and on the Channel Islands it occurs on sand dunes. It overlaps with *E. panzeri* in warm sheltered locations on some sea cliffs. Its occurrence in Wales is limited to the Gower peninsula where it has been found at more than ten separate locations around the coast. Rather more surprisingly it was found in 1988 at Wangford Carr, a small woodland on the edge of a valley fen in Breckland, and in 1989 at Charley Wood, Kirkby, near Liverpool. Intriguingly, there is an old, unlocalised record for West Norfolk, and Lucas (1920) recorded it as a casual through Birkenhead Docks. These recent outlying occurrences suggest that it may yet be found more widely in southern and central England.

Ectobius panzeri Stephens, 1835

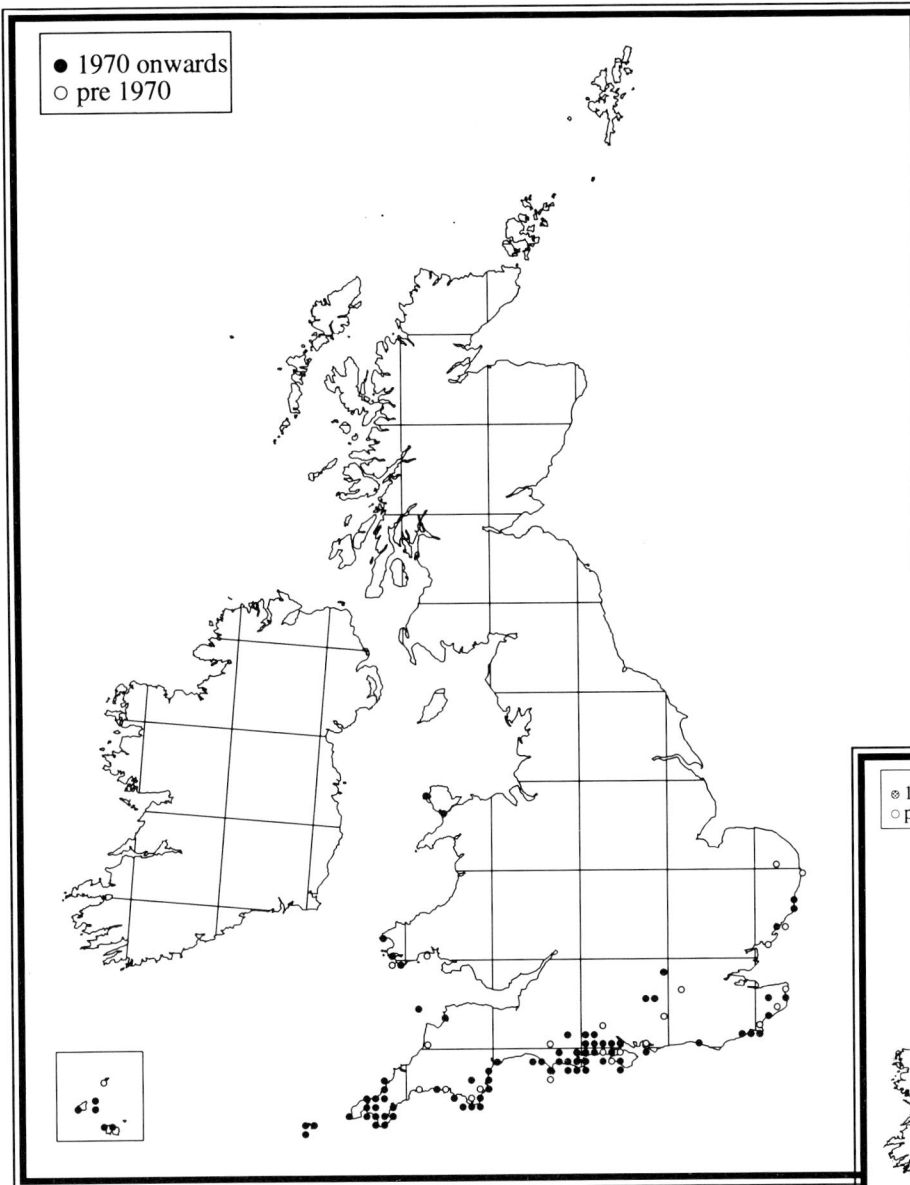

- 1970 onwards
- pre 1970

Marshall and Haes (1988) Description and species account pp131–132, Plate 9.

GB protection/threat status Nationally Scarce (B).

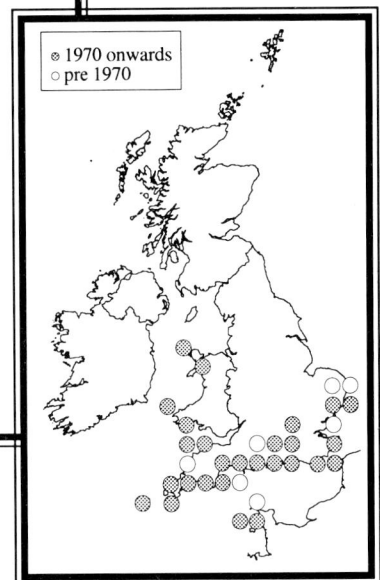

- 1970 onwards
- pre 1970

Status

Native in England and Wales, mainly in coastal areas and in some heath and chalk areas in southern England. Occurs on all the larger Channel Islands, the Isles of Scilly, Lundy and Anglesey.

This, the smallest native cockroach (5–8 mm), is normally dark brown with some speckled patterning on the pronotum in both sexes, and in females also on the dorsal surface of the abdomen, but there is some variation in colour. Males are fully winged and fly well in warm weather, but females have truncated forewings and vestigial hindwings and cannot fly. Unlike the other two species, *E. panzeri* has only a one-year life cycle. Eggs hatch in late April or May and the nymphs, which have distinctive white patterns on the thorax, pass through five instars to mature between July and August, with adults present until late September or early October. It is mainly a coastal species, occurring on sea cliffs, sand dunes and shingle beaches, but during the last 20 years there seems to have been an increase in the number of inland records - from dry heathland, chalk grassland and occasionally woodland. In Wales it has been recorded only in the extreme west: in Pembrokeshire and on Anglesey, where its most north-western occurrence is on South Stack at Holyhead. In England, Dorset and West Cornwall are the main centres of its distribution and it is also widespread in the New Forest. In the east, records are rather scattered and isolated and it has been recorded infrequently in East Anglia, recently at Thorpeness and Minsmere in Suffolk. Recent inland records from Berkshire and Buckinghamshire are of solitary individuals found in situations suggesting that they may have been introduced accidentally.

Dermaptera – earwigs

Earwigs are among the few insects that members of the general public recognise instantly. A record of 'an earwig' in Britain or Ireland is likely to refer to only one species, *Forficula auricularia*, which is almost certainly ubiquitous throughout these islands. This, the commonest species, is a remarkably successful animal, being able to survive wherever there are shelter and maternal nest sites and an adequate supply of food (which could be almost any plant or animal matter). Especially in the autumn, adult earwigs can turn up anywhere – in gardens, parks, waste ground, in hedgerows, among leaf litter and inside buildings.

However, eight species of earwig have been reliably recorded with breeding colonies in Britain and three species in Ireland, of which four in Britain and two in Ireland are regarded as native. The native species are mapped and described here (see pp47–51). The introduced species are very rare and, with one exception, have not been seen for some years. They are described briefly below. For more details, see Marshall and Haes (1988).

Euborellia annulipes (Lucas, 1847) Ring-legged earwig (p134, Plate 9) The only recent record is of a specimen from a house in Warley, Staffordshire, in 1991, which was probably introduced with imported sea shells.

Anisolabis maritima (Bonelli [in Géné], 1832) Maritime earwig (Not described or figured by Marshall & Haes (1988)) Believed to have established temporarily in 1856 at South Shields, Co. Durham.

Maravia arachidis (Yersin, 1860) Bone-house earwig (pp135–136, Plate 9) Has not been recorded since the 1950s.

Labidura riparia (Pallas, 1773) Giant or tawny earwig (p139, Plate 9) Most records were from the coast in the Christchurch and Bournemouth area of Hampshire and Dorset, from 1808 up to 1919 and possibly to the 1930s, and from Folkestone, Kent, in the 1870s. There is a tantalising 'might have been' record from the Seaton area of South Devon in 1985, which cannot now be substantiated. Possibly native on the south coast but, like *Pseudomogoplistes squamiger*, it is more probably an accidental introduction which became established for over a century.

Distribution of records of native earwigs

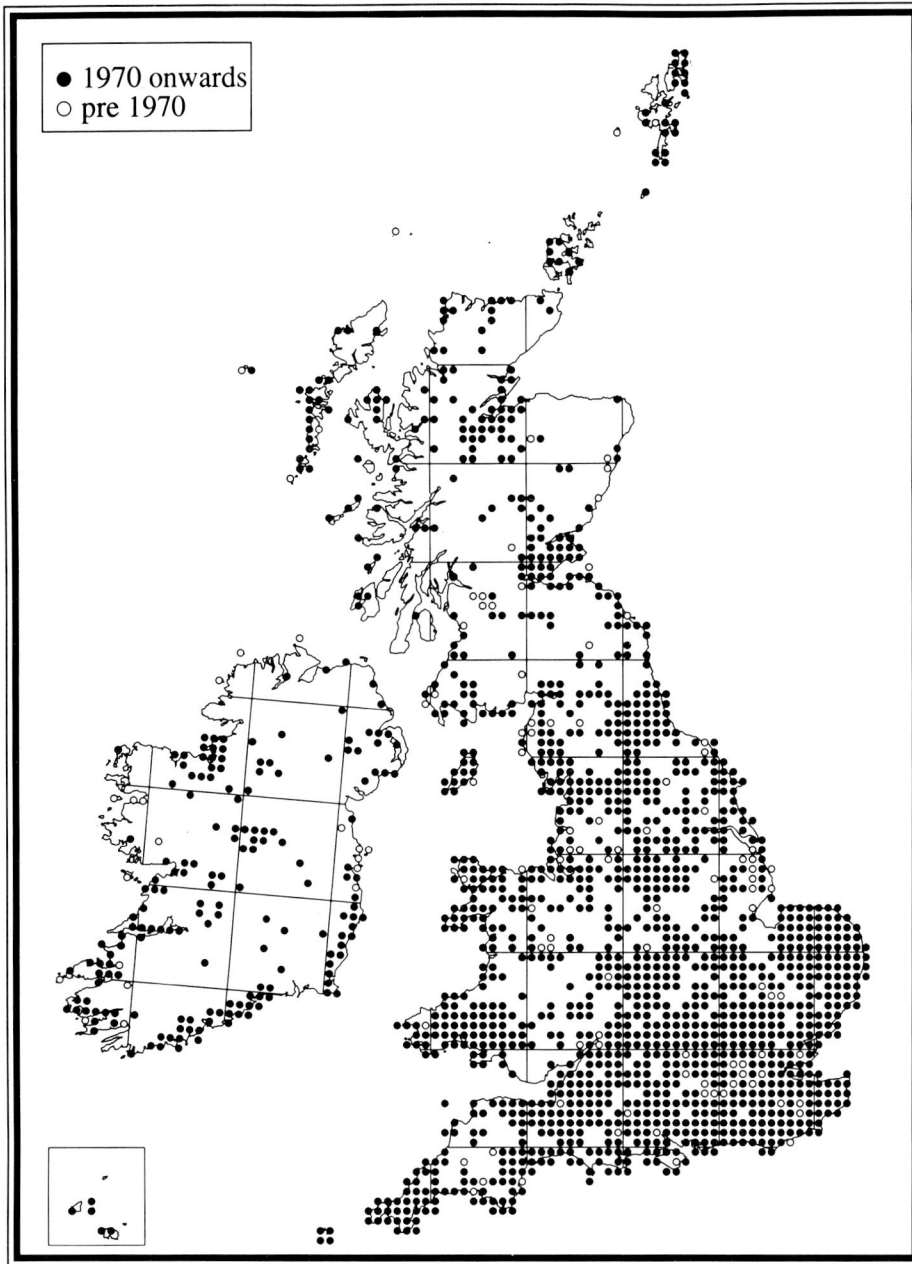

Species accounts and distribution maps

Labia minor (Linnaeus, 1758)

Lesser earwig

Marshall and Haes (1988) Description and species account p135, Plate 9.

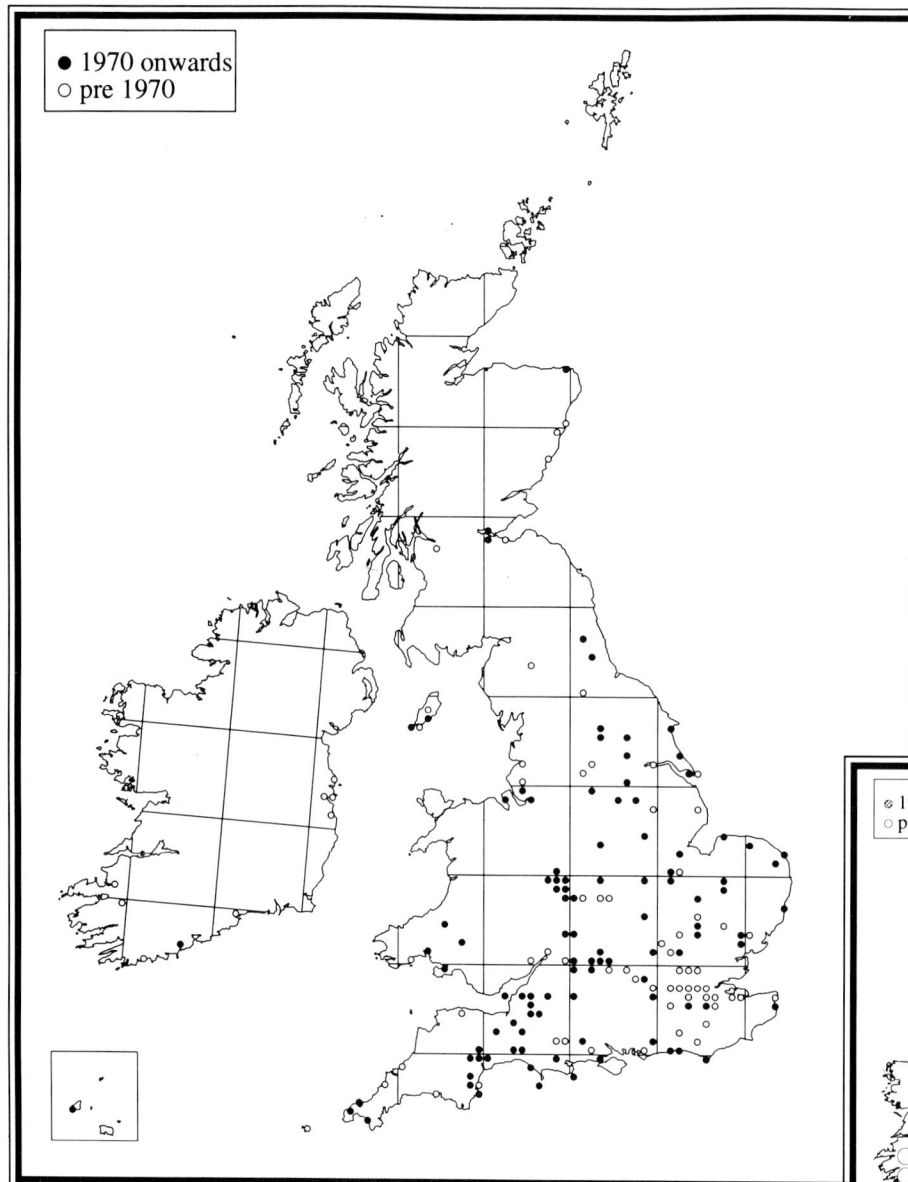

● 1970 onwards
○ pre 1970

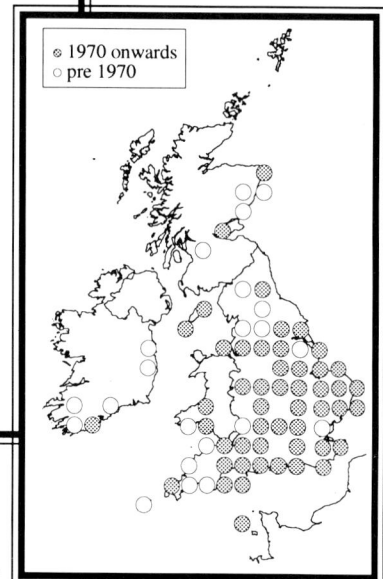

⊘ 1970 onwards
○ pre 1970

Status

A native or a long-established naturalised species in Britain and Ireland. It has almost certainly been overlooked and under-recorded throughout its range. The map is based on localised records, but there are many more early records, for example from Ireland and Scotland, for which localised information is unavailable.

This is the smallest European earwig (4–6 mm), which has been likened superficially to a staphylinid (rove) beetle, although the presence of forceps easily distinguishes it as an earwig. It is a dull, yellowish brown, with a pubescent body and dark head. Both sexes are fully winged and fly readily, being active both in sunshine and at night, and are attracted to light. This species is closely associated with dung, compost and rubbish heaps in rural or urban areas – anywhere there is consistent, moist, natural heating from decaying organic matter. In such conditions, breeding and development can occur throughout the year, so all stages may occur together. It has a life cycle of about three months. The abundance and spread of early records (mostly from before 1950) suggest that this species may have become less common in recent decades. This is not unlikely with the decline in numbers of stabled livestock (especially horses) and resultant long-term dung heaps, and also more recently with the use of broad-spectrum, veterinary, pharmaceutical chemicals to control internal parasites in livestock, as some of these chemicals persist in dung. However, there have been several records since 1990 from dung heaps, or at lepidopterists' light traps, at locations from South Devon to north-east Scotland.

Apterygida media (Hagenbach, 1822) **Short-winged or hop-garden earwig**

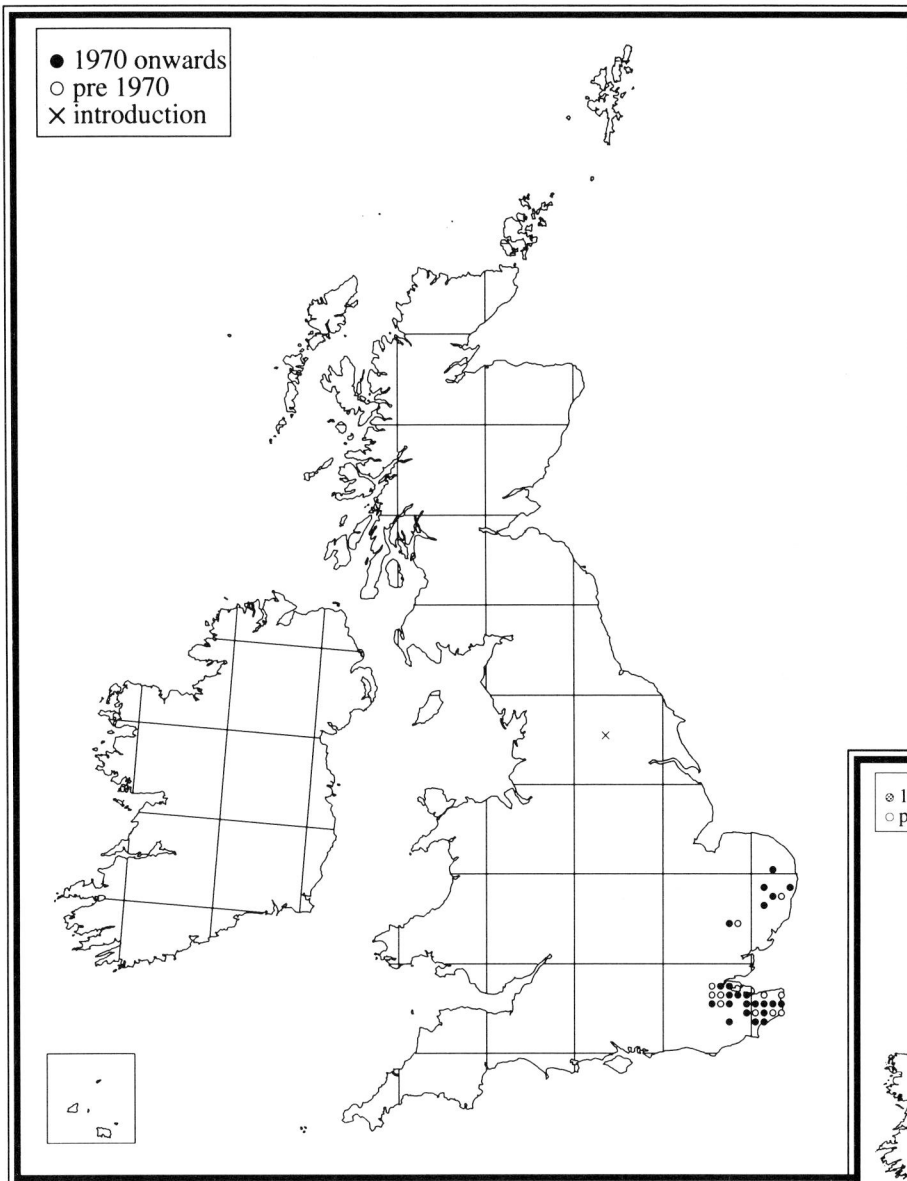

● 1970 onwards
○ pre 1970
✕ introduction

⊗ 1970 onwards
○ pre 1970

**Marshall and Haes
(1988)** Description
and species account
p136, Plate 9.

**GB protection/threat
status** Nationally
Scarce (B).

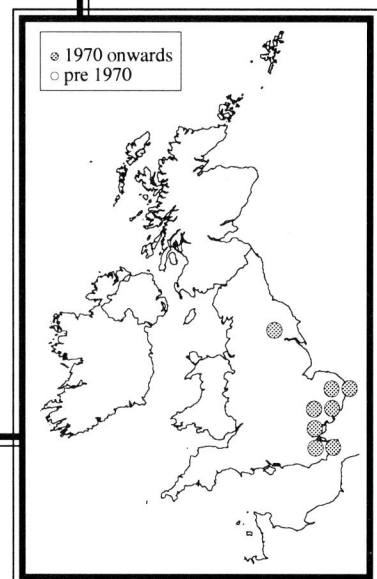

Status

Probably native in the drier and warmer parts of south-east England.

This species is reddish brown with yellow legs and is intermediate in size (6–10 mm) between the common earwig and the other two native species. Females are very similar in appearance to *Forficula lesnei*. The hindwings are only tiny lobes and consequently it is flightless. Little is known of its life cycle or biology. Adults are usually present in August and September and are nocturnal. They appear to hibernate, usually under bark or among moss. This was formerly a common insect in the hop-gardens of Kent, but seems to have declined since the widespread use of insecticides and mechanical picking of hops have been prevalent. It now occurs in small numbers on trees and shrubs, including the edges of woods, hedges and domestic gardens. It has been found once at the bottom of a cliff in Suffolk. Although hops were, and in some areas still are, grown in other areas of England (eg Herefordshire, Worcestershire and Hampshire), there are no records of *A. media* from these areas.

Forficula auricularia (Linnaeus, 1758)

Common earwig

Marshall and Haes (1988) Description and species account pp137–138, Plate 9.

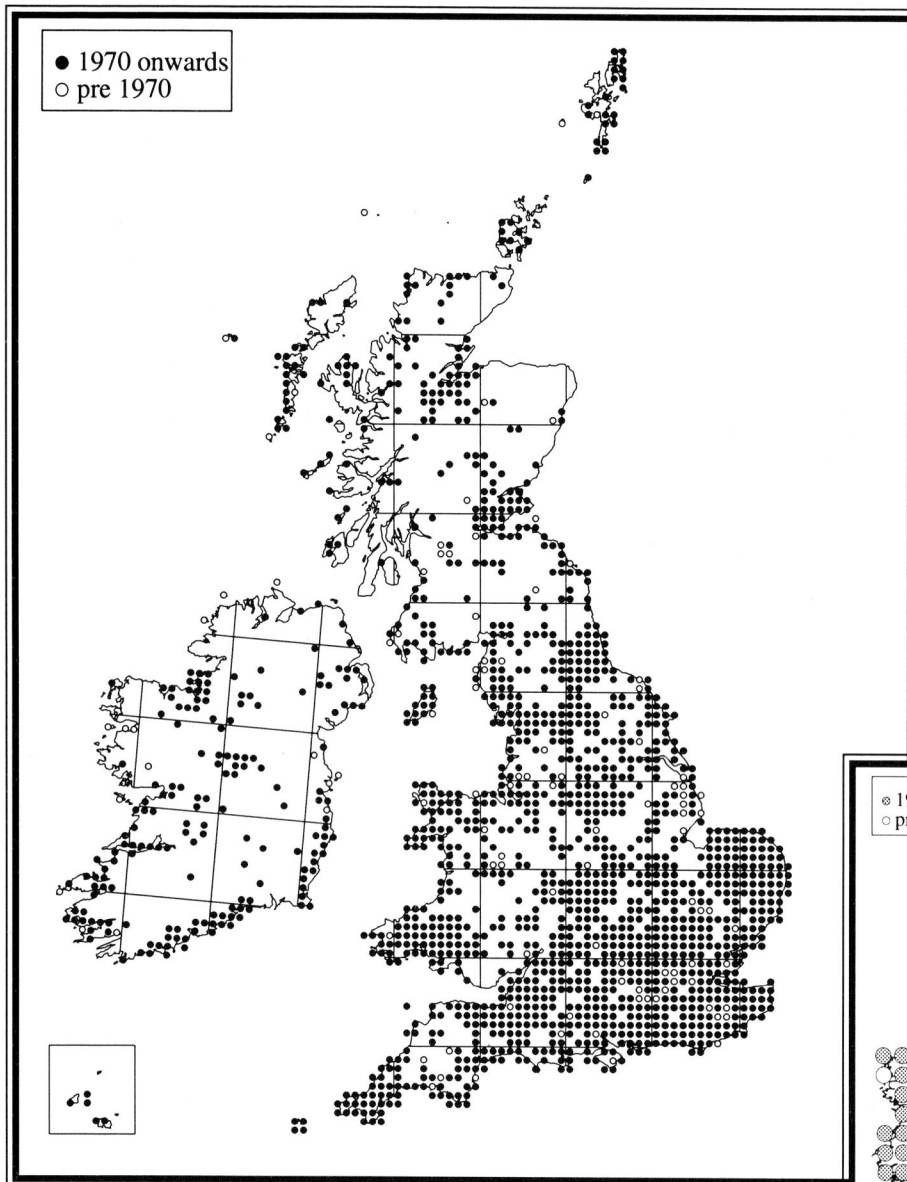

● 1970 onwards
○ pre 1970

⊗ 1970 onwards
○ pre 1970

Status

Very widely distributed and generally common throughout Britain and Ireland, including many offshore islands. Probably one of our commonest, most obvious and familiar large insects, and, because it is so common and obvious, recorders have often not bothered to record it. The map is misleading and in many areas reflects the distribution of diligent recorders rather than the insect itself.

This shiny, dark chestnut-brown earwig varies in size from 10 to 15 mm. The forceps are variable and a distinctive large form *forcipata* occurs, particularly on offshore islands. Although the common earwig is fully winged, it is rarely observed in flight, but it is normally active at night. The life history has been studied extensively, and includes complex maternal care of eggs and nymphs through the winter and spring. It seems there are four nymphal instars, though continental populations have five. Adults can be observed at most times of the year, but are most noticeable in the autumn, when they may commonly be shaken out of dahlias and chrysanthemums in gardens. Females are capable of rearing two broods in a 12-month period. It occurs most typically among long vegetation, under bark and among other debris in almost any habitat. It is less frequent in open, dry locations such as arable fields, short grazed or mown turf and dry heath, and in saltmarshes. It has been found to be common on many islands including the Outer Hebrides (where it is common in machair grassland), St Kilda and the Shetland Islands.

Forficula lesnei (Finot, 1887)

Lesne's earwig

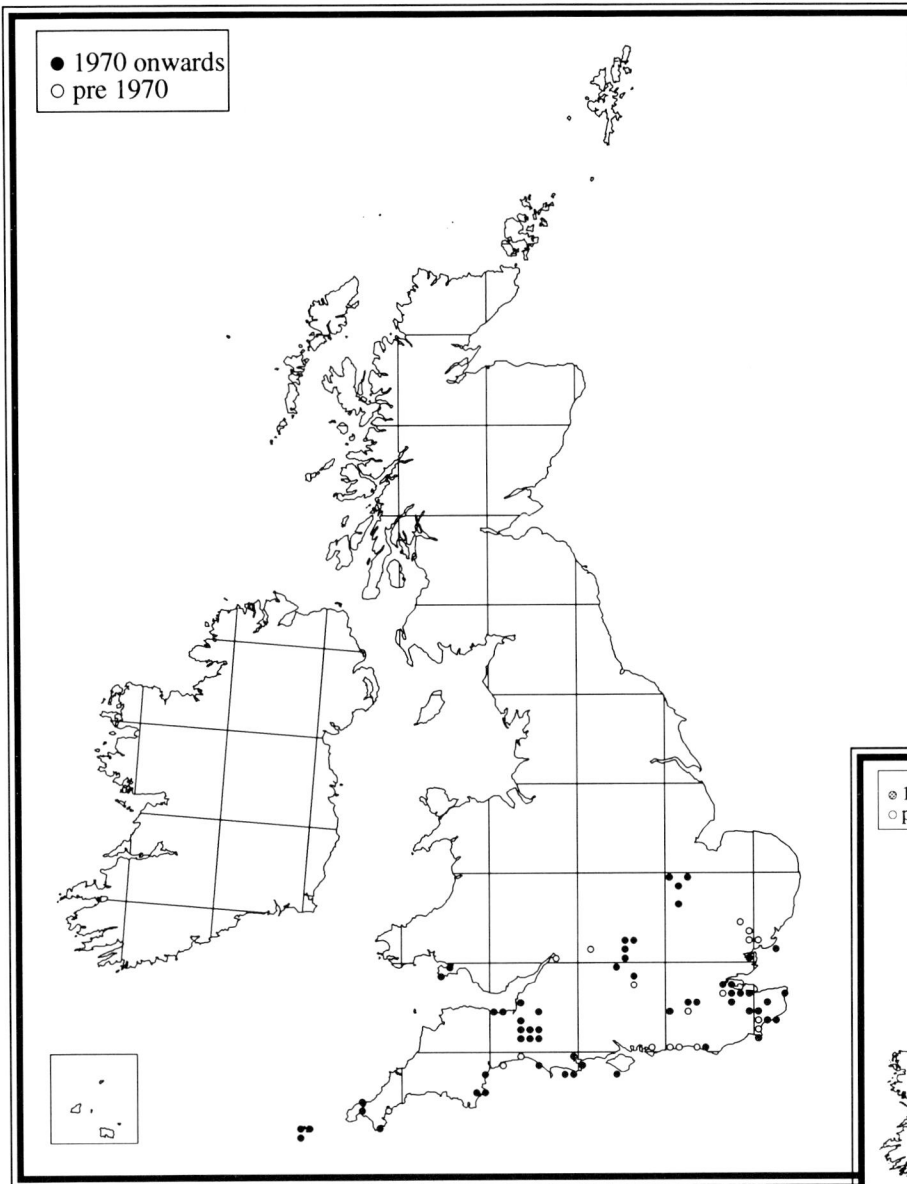

Marshall and Haes (1988) Description and species account p138, Plate 9.

GB protection/threat status Nationally Scarce (B).

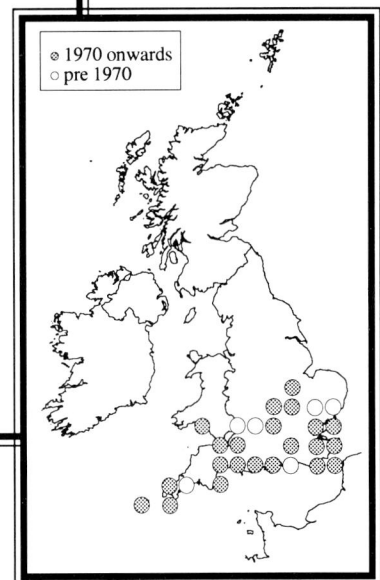

● 1970 onwards
○ pre 1970

⊘ 1970 onwards
○ pre 1970

Status

Recorded from scattered locations in southern England and on the Gower peninsula in south Wales. Recently rediscovered in the Isles of Scilly. An elusive species and almost certainly under-recorded. On the northern edge of its range in Britain and may be restricted to particularly favourable locations, which have not yet been characterised.

This is a small species (6–7 mm) somewhat like a pale, small version of the common earwig, but the hindwings are either absent or greatly reduced. Aspects of its life history, particularly parental behaviour, have been investigated recently (Timmins 1995). It is normally active at night. Although adults and nymphs have been recorded as early as May, most records of adults have been from July to October. There is little pattern to its occurrence, although most recent records have been from sites on base-rich soils. It has been beaten from trees and shrubs, and has been recorded from hedges, among nettles (*Urtica* spp) and in rough vegetation, including that on sea cliffs, and in particular inside hollow stems of hogweed (*Heracleum sphondylium*) and other umbelliferous plants.

51

Phasmida – stick-insects

More than ten species of phasmid are known to occur in continental Europe. Stick-insects are popular 'pets' and there is an active Phasmid Study Group. It is known that accidental escapes and intentional releases of surplus stock occur, leading to occasional outdoor finds of the more commonly reared species. Most 'pet' species are, however, totally unsuited to our climate and do not survive long. Nevertheless, at least five kinds of stick-insect have been recorded regularly in Britain, and one in Ireland, of which three appear to be established as breeding populations outdoors in very mild, damp areas. The taxonomic status of four of these taxa has been reviewed since the publication of Marshall and Haes (1988).

Carausius morosus (Sinéty 1901) Laboratory (or Indian) stick-insect
Marshall and Haes (1988) pp143-144, Plate 10.
This is a distinctive species and adults are considerably smaller than the other stick-insects recorded regularly in Britain and Ireland. It is oriental in origin (from southern India) and does not sustain breeding populations here, other than in glasshouses. There were colonies for many years centred on glasshouses at Torquay, Devon, and Kew Gardens, London, but both have died out. Recently, a further, well-established population has been found at a butterfly centre in the Midlands.

Three of the other taxa (two sub-species of *Acanthoxyla prasina*, and *Clitarchus hookeri*) originated in New Zealand and are apparently naturalised here in the extreme south-west of England and one species in south-west Ireland. However, there had been considerable confusion about their taxonomic status and true identity, which have been clarified recently by Brock (1987, 1991). The genus *Acanthoxyla* has been revised by Salmon (1992) and two of the supposed species found in Britain and Ireland are now considered to be sub-species of *A. prasina*, the black spined stick-insect. The origins of these populations are considered by Lee (1995): they were originally found in localities where there are landscaped gardens and nursery gardens containing collections of plants originating from New Zealand, Tasmania and Australia. Antipodean species from other taxonomic groups (eg terrestrial flatworms and the amphipod *Arcitalitrus dorrieni* (Hunt) (the so-called wood-hopper)) are known to have become established at several of the same sites. It is perhaps surprising that, although stocks of living plants from New Zealand have been widely distributed in the temperate zone of the northern hemisphere, these phasmids appear to have become established only in Britain and Ireland.

Acanthoxyla prasina geisovii (Kaup, 1866) Prickly stick-insect
Marshall and Haes (1988) pp140-141, Plate 10.
Recorded from the Isles of Scilly, Cornwall and Devon, mainly from gardens, it may have been present at Paignton Zoo, Devon, since 1908 and possibly at Tresco Abbey, on the Isles of Scilly, since 1907.

Acanthoxyla prasina inermis Salmon, 1955
Unarmed stick-insect
Marshall and Haes (1988) pp142, Plate 10
Although not recognised in Britain until 1987 because it was confused with *C. hookeri*, there is circumstantial evidence that *A. prasina inermis* has been present in the Falmouth and Truro districts of Cornwall since the 1920s. There was a further introduction to Cornwall through a nursery in Somerset in the 1970s, so that two periods of introduction have occurred. Since the late 1980s, it has been known from a cluster of sites along the north coast of Cornwall (P D Brock pers. comm. and Lee 1995). Brock and Lee (1995) have confirmed that the stick-insects recorded in the Kenmare River area of south-west Ireland are this species and not *C. hookeri* as previously assumed.

Clitarchus hookeri (White, 1846) Smooth stick-insect
Marshall and Haes (1988) p143, Plate 10
Known with certainty only from Tresco, Isles of Scilly.

Bacillus rossius (Rossi, 1788) Corsican stick-insect
There are records of this species from garden hedges at St Mawes, Cornwall, in 1992 and Plympton, Devon, in September and October 1994 (Lee 1995). A single, adult female was found in July 1995 at a garden in Hillingdon (north London) resulting from a deliberate introduction made at the same site in 1986 (McNamara 1996).

Sipyloidea sipylus (Westwood, 1859) Pink-winged stick-insect
McNamara (1996) also implies other deliberate releases were made of exotic stick-insects at the garden in Hillingdon and at nearby Pole Hill Wood in 1986, including *C. morosus* (q.v.) and *S. sipylus*, with the latter surviving to the following summer.

Distribution of records of naturalised stick-insects

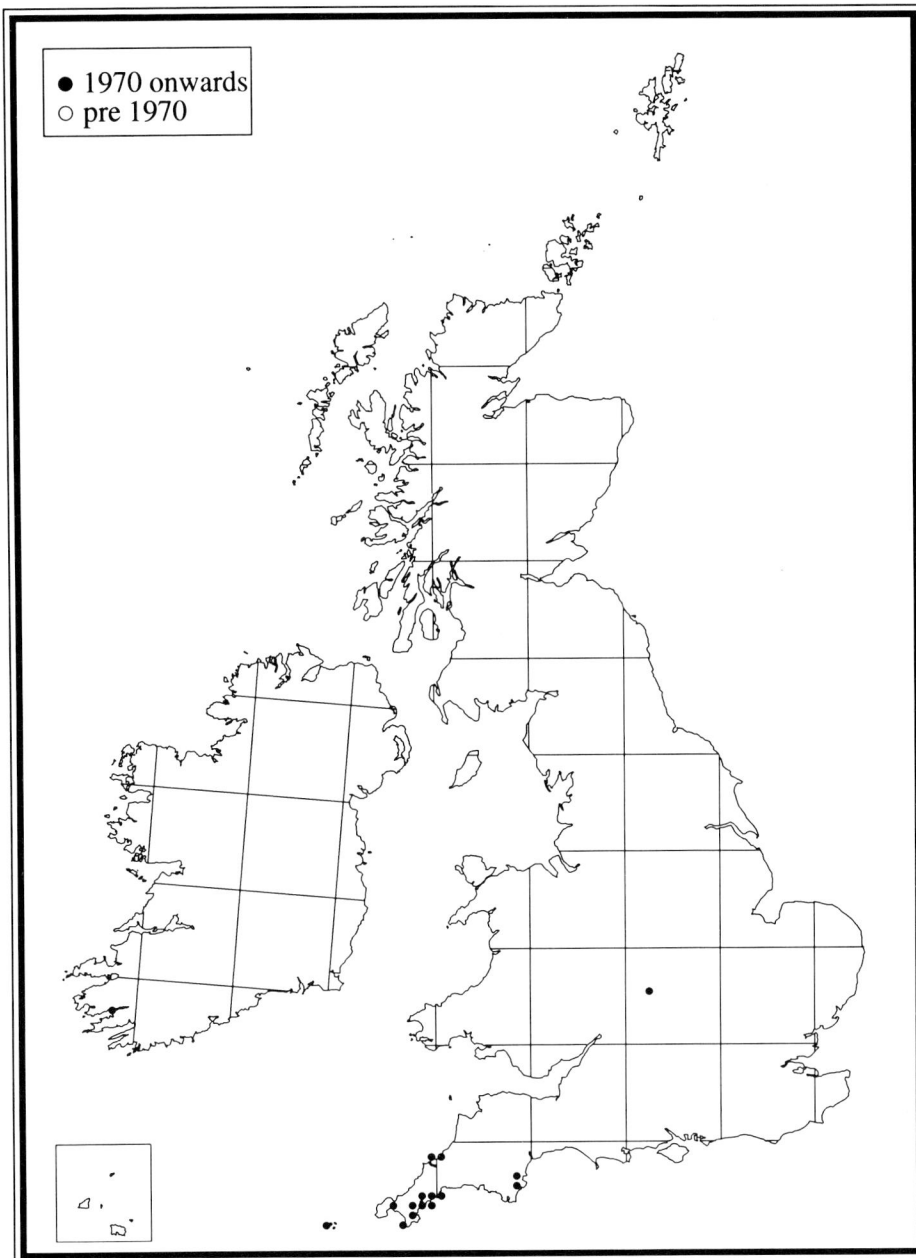

● 1970 onwards
○ pre 1970

53

Acknowledgements

Apart from the contributors of records singled out for special mention in the **Introduction**, we are delighted to acknowledge the contribution made by the many individual recorders (see below) who have contributed to the scheme.

We should also like to thank Brian Eversham and Henry Arnold for their work with the scheme at the Biological Records Centre and with preparing the data and maps for this *Atlas*. Previous Data Managers at BRC (particularly the late Dorothy Greene and also Claire Appleby and Julian Dring) also contributed to work on the database over the years. Most of the keyboarding of data from record cards and the initial editing of data were undertaken at BRC by Val Burton and Wendy Forrest. Also, we should not forget the late John Heath who set up the original recording scheme and Michael Skelton who helped John to run the scheme in its early years.

New information and help with revision of the accounts of the following species were kindly provided by Dr Andrew Cherrill (*Decticus verrucivorus* and *Stenobothrus stigmaticus*) and Mike Edwards (*Gryllus campestris* and *Gryllotalpa gryllotalpa*). We are also grateful to Dr Stuart Ball and Professor Val Brown for helpful comments and additions to the text.

The maps were produced using the DMAP program written by Dr Alan Morton.

Finally, we acknowledge the contribution made by Penny Ward and Karen Goodsir in preparing our copy for publication.

The Recorders

The following list includes the recorders whose data are included in this *Atlas*. Most have supplied records directly to the recording scheme, some are responsible for published records, and others for specimens in museum collections. Records from the literature and collections provided a valuable historical perspective.

Abraham, J.; Acuna, T.; Adams, J.H.; Adams, R.G.; Addey, J.E.; Airy-Shaw, H.K.; Albertini, M.V.; Alexander, K.N.A.; Allan, J.; Allen, C.; Allen, D.; Allen, G.W.; Allen, N.; Allwright, A.R.; Almond, W.E.; Amphlett, M.; Anderson, M.; Anderson, R.; Andrewes, C.H.; Andrews, M.; Angus, A.; Angus, S.; Anson, A.E.; Appleby, M.;

Appleby, S.; Appleton, D.; Appleyard, A.; Archer, M.W.; Arnold, G.A.; Arnold, H.R.; Arnold, M.A.; Arthur, R.W.; Ash, H.J.; Ashby, E.B.; Ashby, G.T.; Ashby, S.R.; Ashton, H.F.; Askew, R.; Atkins, R.; Atkinson, E.H.; Atkinson, M.; Atty, D.; August, J.E.; Austin, M.; Austin, R.; Aynsley, J.; Babington, C.; Baccus, M.; Badmin, J.S.; Bagnell, H.S.; Bailey, K.; Bailey, T.; Bainbridge Fletcher, T.; Baird, B.; Baker, B.R.; Baker, P.F.; Baldock, D.W.; Baldwin, M.C.; Ballantyne, G.H.; Banks, J.L.C.; Banks, R.; Barker, M.; Barker, S.; Barks, R.; Barnes, A.E.; Barnes, R.M.; Barnett, S.L.; Barnham, M.; Barroa, N.; Barrowcliffe, K.; Batchelor, D.M.; Bate, G.E.; Bateman, R.D.; Bateman, R.P.; Bathe, G.M.; Bayford, E.G.; Bedwell, E.C.; Beith, G.; Bellamy, G.; Benham, B.R.; Benson, R.B.; Benton, E.; Berks, Bucks and Oxon Naturalists Trust; Berman, H.J.; Biglin, J.; Bignell, G.C.; Billups, C.R.; Bind, R.; Binding, A.E.; Bindon, C.; Birchenhall, D.; Birkenhead, G.A.; Birkett, H.L.; Birkett, N.L.; Birt, Mrs; Bishop, A.H.; Bishop, E.; Bishop, G.W.; Blackburn, P.; Blackith, R.E.; Blackman, A.W.; Blackman, T.J.; Blackwell, C.; Blackwell, K.; Blackwell, R.; Blair, K.G.; Blane, E.; Blofeld, V.; Bloomfield, E.M.; Boardman, E.T.; Bodycote, M.; Boers, B.; Bolton, A.; Bolton, D.E.; Bond, F.; Boniface, A.; Booth, F.; Booth, J.S.; Boreham, L.R.; Boswell, H.; Boulton, C.; Bourne, A.C.; Bourne, R.A.; Bowdrey, J.; Bowell, M.; Bowen, H.J.M.; Bowers, J.; Bowman, N.; Box, H.E.; Boyce, D.C.; Boyer, J.A.J.; Bradford, E.S.; Brassley, P.; Bratton, J.H.; Bray, R.P.; Breasley, S.; Breeds, J.; Brewster, M.A.; Brian, M.; Bridson, M.; Briggs, C.A.; Brind, R.A.; Brindle, A.; Britten, H.; Britton, E.B.; Brock, P.D.; Broodbank, A.R.; Brook, J.; Brookhuyson, G.J.; Broomfield, E.N.; Broomfield, P.S.; Broughton, W.B.; Brown, A.J.; Brown, E.; Brown, E.R.; Brown, E.S.; Brown, G.; Brown, H.; Brown, J.M.; Brown, P.; Brown, S.C.S.; Brown, V.K.; Brownsword, C.; Brownsword, M.; Brummitt, J.M.; Bryant, A.; Buchanan, J.; Buck, F.D.; Buck, M.E.; Buckingham, D.; Buckley, J.; Buckley, W.; Bull, A.; Bull, A.L.; Bull, C.M.; Bulley, H.; Bullock, E.F.; Bullock, I.; Bundy, A.; Bunn, D.S.; Bunting, W.; Burke, F.; Burr, M.; Burton, H.; Burton, J.F.; Burton, P.A.; Burtt, E.; Butcher, P.; Butler, C.G.; Butler, H.; Butler, N.D.; Butlin, R.K.; Buxton, P.A.; Bysouth, E.B.; Campbell, G.; Campbell, I.K.; Campbell, J.I.; Campbell, J.M.; Campbell, S.; Campbell, W.D.; Campion, F.W.; Campion, H.; Cannon, J.F.M.; Cant, A.; Carlisle, N.H.S.; Carney, K.; Carpenter, G.H.; Carstairs, D.; Carter, D.J.; Carter, H.H.; Carter, J.W.; Carter, W.A.; Catt, M.; Cawley, M.; Chalmers-Hunt, J.M.; Cham, S.; Chamberlain, A.;

Champion, G.C.; Champion, H.; Chapman, D.I.; Chapman, F.I.; Chapman, F.J.; Chapman, M.; Chapman, R.A.; Chardle, P.A.; Charlson, S.; Chater, A.O.; Cheetham, C.A.; Chelmick, D.G.; Chenery, J.M.; Cherril, A.J.; Chick, D.H.; China, W.E.; Chinery, M.; Chinnery, G.A.; Chitty, A.J.; Church, A.R.; Cinchie, D.J.; Claridge, M.F.; Clark, A.; Clark, D.J.; Clark, E.J.; Clark, J.; Clark, P.; Clarke, J.; Clarke, K.V.; Clarke, R.; Clarkson Webb, C.M.; Classey, E.W.; Clatworthy, R.; Cleave, A.; Clegg, T.M.; Clements, D.K.; Clements, H.A.B.; Clements, K.M.; Clemons, L.; Clennett, D.J.; Clinging, R.; Cloudsley-Thompson, J.L.; Clynes, W.; Clynes, W.W.; Coats, J.S.; Cobb, P.; Cockerell, T.D.A.; Cocks, W.P.; Coe, R.L.; Coetzee, E.F.; Cogan, B.H.; Cohen, E.; Cole, C.B.; Coles, A.; Coley, M.; Collier, R.V.; Collins, G.B.; Comins, J.; Comont, S.J.; Condry, W.M.; Conolly, A.; Cook, I.; Cook, K.; Cooke, K.; Cooke, P.; Coombes, F.W.; Cooper, D.; Cooper, J.; Copson, P.J.; Corbett, H.H.; Corley, M.; Cornwell, P.B.; Cotteswold Naturalists Field Club; Cotton, D.C.F.; Cotton, M.J.; Cove, W.; Cowin, W.S.; Cowley, J.; Cox, J.R.; Crane, M.; Craske, R.M.; Crawshaw, K.R.; Creed, P.; Crocker, J.; Crockett, B.G.; Cropper, R.S.; Crosby, T.S.; Cross, E.; Cross, I.; Cross, S.N.; Crosskey, M.A.; Crosskey, R.W.; Crowson, R.A.; Cumber, R.A.; Cumming, R.; Currie, P.W.B.; Curtis, R.J.; Dack, C.; Dackus, T.; Dale, C.W.; Dale, J.C.; Daltry, H.W.; Danby, C.R.; Daniel, M.C.; Daniels, E.T.; Dapling, J.; Darley, P.; Davey, S.R.; David, C.T.; Davidson, C.; Davies, D.; Davies, I.P.; Davies, M.; Davis, J.; Davis, M.J.; Davis, T.A.W.; Davis, W.; Daws, J.; Dawson, G.; Dawson, J.E.; Day, C.D.; Day, F.H.; Day, G.M.; Dean, M.; Dean, W.F.; Dear, L.; Debbage, P.; Dempsey, M.; Dennis, G.; Denton, J.; Devriese, H.; Dewhurst, C.F.; Diamond, H.; Diamond, M.; Dibb, J.R.; Dibbs, D.J; Dicker, G.H.L.; Dickson, R.J.; Dingle, T.; Distant, W.L.; Diver C.; Diver, P.; Dobson, J.; Dobson, V.; Dolling, J.; Dolling, M.; Dolling, W.R.; Doncaster, J.P.; Donisthorpe, H. St J.; Donovan, J.; Donovan, T.W.; Dorling, D.A.; Dorset Environmental Records Centre; Dorset Trust for Nature Conservation; Doughty, C.G.; Douglas, J.M.; Down, D.; Drake, M.; Draper, F.; Driscoll, P.A.; Driscoll, R.J.; Drost, W.; Du Feu, G.R.; Ducker, S.; Dudley, L.C.; Duff, A.; Duffy, E.; Duncan, J.; Duncan, Sir A.; Durnell, P.R.; Durrant, K.C.; Dutson, G.; Eagles, T.R.; Earl of Cranbrook; Eccles, T.; Eden, S.; Edgell, F.B.; Edgington, M.; Edwards, B.; Edwards, D.; Edwards, J.; Edwards, M.; Element, D.; Elias, D.O.; Elliott, M.; Elliott, R.; Ellis, A.E.; Ellis, E.A.; Ellis, J.; Ellis, J.W.; Ellison, N.F.; Else, G.R.; Elton, C.; Elton, D.; Ely, W.A.; Embry, B.; Emley, D.; Enfield, M.; English, P.; Epping Forest Conservation Centre; Evans, D.E.; Evans, I.M.; Evans, R.; Evans, S.B.; Evans, T.; Eve, H.C.; Eversham, B.C.; Everyweek Survey; Eyles, E.; Fairclough, K.; Fairhurst, M.; Farris, R.C.; Farrow, F.; Farrow, R.A.; Farthing, J.; Felton, J.C.;

Fennell, C.V.; Fergusson, N.M.P.; Ferroussat, F.; Fieldhouse, D.S.; Fielding, E.H.; Finch, R.A.; Fincher, F.; Finlow, B.; Finnemore, M.W.; Fitton, M.G.; Fleet, Susan; Fletcher, W.H.; Flint, J.H.; Flumm, D.S.; Fogan, M.; Fonseca, E.C.M.d'Assis; Foord, R.G.; Foord, S.C.; Ford, B.; Ford, P.L.; Ford, R.; Ford, R.J.; Ford, S.C.; Ford, W.K.; Fordham, W.J.; Forrester, L.; Foster, A.P.; Foster, G.N.; Foster, T.Le Neve; Fowler, W.W.; Fowles, A.P.; Fowling, R.; Fox Wilson, G.; Fox, A.D.; Fox, D.P.; Foxwell, D.J.; Francis, I.; Frankum, M.; Fraser, F.C.; Fraser, W.R.; Fray, P.J.; Freeman, P.; Freeman, R.; French, C.N.; French, W.L.; Friedlander, C.P.; Frisby, G.E.; Frost, R.A.; Fry, R.M.; Fylde Naturalists; Gamble, P.H.; Gandy, M.; Gardiner, C.J.; Gardner, A.E.; Garland, S.; Garrad, L.S.; Garretty; Gatward, N.M.; Geiger, G.; Geiger, P.; George, R.S.; Gerrard, P.C.; Gibbs, D.; Gibbs, R.; Gibson, C.; Giddens, C.; Gilbert, O.L.; Gillham, M.; Gilmour, E.F.; Goddard, D.G.; Goddard, P.; Goldsmith, J.G.; Goldthorp, M.; Good, J.A.; Goodall, A.; Goodings, B.M.; Goodley, T.; Goodlife, F.W.; Goodliffe, F.D.; Goodson, A.L.; Goss, I.; Gough, E.; Gowers, S.; Gowing-Scopes, E.; Graham, A.; Graham, M.W.; Grant, J.A.; Granville, A.G.; Graves, P.P.; Grayson, A.; Greathead, D.J.; Green, E.E.; Green, J.; Green, J.A.; Green, S.; Green, S.G.; Green, S.V.; Green, V.A.; Greenslade, R.M.; Greet, R.; Gregory, S.; Grensted, L.W.; Grey, O.; Grice, M.; Griffith, E.M.; Griffiths, P.; Grimes, E.; Grove, S.J.; Guermonprez, H.; Guichard, K.M.; Gush, G.H.; Guthrie, G.P.; Hadley, M.; Haes, E.C.M.; Haigh, D.J.R.; Haines, C.W.; Haines, F.; Hainsworth, P.H.; Halbert, J.N.; Hale Carpenter, J.D.; Halfpenny, G.; Hall, C.; Hall, M.R.; Hall, R.; Hall-Smith, D.; Hallet, H.M.; Halls, J.; Hammond, C.O.; Hammond, H.E.; Hammond, R.; Hampshire and Isle of Wight Wildlife Trust; Hance, B.E.A.; Hancock, E.G.; Hancox, C.R.; Harcsa, J.; Harding, P.T.; Hardman, J.A.; Hardy's School, Dorchester; Hargreaves, E.; Harley, A.; Harley, B.H.; Harman, A.; Harper, K.G.; Harpur Crewe, H.; Harris, J.; Harrison, M.G.; Harrogate and District Naturalists Society; Hart, N.; Hartham, B.; Hartley, J.C.; Harvey, D.H.; Harvey, D.J.; Harvey, P.; Harwood, P.; Harwood, W.; Hasnip, C.N.; Hassall, M.; Hawkins, C.N.; Hawkins, R.D.; Hawkins, S.; Hawley, S.; Hay, G.; Hayes, E.; Hayhow, S.J.; Haynes, J.; Haynes, P.; Hayward, J.; Haywood, R.; Heal, N.F.; Heath, P.; Heath, V.F.; Henderson, A.; Hendry, C.H.; Herbert, M.J.; Hesselgreaves, E.; Hewitt, G.M.; Hewitt, S.M.; Higgins, D.G.F.; Higgins, R.; Higgot, J.B.; Highway, F.G.; Hill, R.; Hincks, W.D.; Hind, S.H.; Hinkins, F.R.; Hinks, W.D.; Hipperson, D.; Hobbs, A.; Hobbs, N.; Hobbs, R.N.; Hodgkinson, L.S.; Hodgson, C.J.; Hogg, P.; Hold, A.I.; Hollis, D.; Holmes, J.W.D.; Holmes, R.A.; Hooper, L.; Hooper, L.A.; Hope, W.; Horacsek, L.; Horne, D.W.; Horne, R.; Horton, G.A.N.; Horton, P.J.; House, S.;

Hubbard, A.; Hughes, M.R.; Hughes, R.A.D.; Humphreys, J.; Humphries, M.; Hunford, D.A.J.; Hunt, D.; Hutchinson, G.; Hutson, A.M.; Huxley, J.; Hyman, P.; Ikin, H.; Inns, H.; Irving, L.; Irwin, A.G.; Irwin, B.J.; Irwin, T.; Ismay, J.W.; Jackson, A.S.F.; Jackson, B.; Jackson, E.; Jackson, N.; Jacoby, M.; Jago, J.; James, T.; Jansen, E.; Jarvis, M.; Jeffreys, D.M.; Jeffreys, J.G.; Jennings, R.; Jepson, P.; Jerrard, P.C.; Jobe, J.B.; John, D.; Johnson, A.; Johnson, C.; Johnson, G.; Johnson, J.; Johnston, A.; Johnston, H.B.; Jolley, D.R.; Jones, A.; Jones, B.; Jones, B.L.; Jones, C.M.; Jones, D.K.; Jones, M.; Jones, N.P.; Jones, P.; Jones, R.; Jones, R.E.; Jones, S.; Jones, S.P.; Jones, W.; Joyce, C.; Judd, S.; Justin, S.H.; Kearney, J.; Keeby, J.; Keen, D.; Keep, M.; Keeping, V.M.; Kefford, R.W.K.; Kelham, A.; Kemp, R.; Kendall, T.; Kennard, A.; Kent, M.; Kenward, H.K.; Kershaw, G.B.; Kesby, J.; Kesteven, W.H.; Kettle, R.; Kevan, D.K.McE; Key, T.; Keys, J.; Kidd, L.N.; Kidd, M.E.; Killingbeck, J.; Killington, F.J.; Kimmins, D.E.; Kirby, P.; Kitchen, A.H.; Kitchen, T.B.; Knight, R.; Knight, S.; Knox, J.; Kruys, I.P.; Lack, D.L.; Lade, J.A.; Laing, F.; Lambert, S.J.; Lancashire, B.; Land, R.; Lane, C.; Lane, S.; Lang, E.; Langdon, R.; Lansbury, I.; Large, A.N.; Last, B.; Latham, L.; Laurence, B.R.; Law, M.A.; Lawman, D.C.; Lawrence, D.; Lawton, J.H.; Lazenby, A.S.; Le Pard, G.; Le Quesne, W.J.; Le Sueur, F.; LeBrocq, P.F.; Lear, N.W.; Lee, C.; Lee, M.; Leech, A.; Leech, M.J.; Leicester Museum and Art Gallery; Lever, P.D.; Levington, R.; Levy, E.T.; Lewington, R.; Lewis, D.C.; Lewis, E.J.; Lewis, J.W.; Lewis, N.; Liford, R.G.; Lightfoot, P.; Limbert, M.; Lindroth, C.H.; Lindsay, C.; Line, J.M.; Lines, K.; Liston, A.D.; Liverpool Museum; Liverpool and Glasgow Salvage Company; Lloyd, D.F.; Lloyd-Evans, L.; Long, M.L.; Long, R.; Longfield, C.; Lorimer, R.I.; Lott, D.A.; Low, A.M.A.; Low, G.; Loyd, E.K.; Lucas, W.J.; Luff, M.L.; Luff, W.A.; Luvoni, A.B.; Lyle, G.T.; Lynch, W.; Lynes, M.; Lynn, G.; MacLagan, D.S.; Macdonald, E.; Macdonnell, L.M.; Mackay, M.; Mackechnie-Jarvis, C.; Mackie, D.W.; Macnair, V.J.; Macnulty, B.J.; Maddrell, H.; Madge, S.C.; Mahon, A.; Maidstone, R.; Mallet, J.; Manchester Museum; Mann, D.J.; Manning, S.A.; Margoschis, R.A.; Marmont, A.M.; Marrable, D.; Marriott, P.; Marrs, B.; Marsay, C.; Marsh, A.; Marsh, G.; Marshall, A.; Marshall, J.A.; Marshall, J.B.; Marshall, R.; Martin, C.; Martin, D.; Martin, J.; Martin, N.A.; Marton, D.S.; Maskrey, J.; Mason, J.L.; Mason, J.M.; Mason, R.L.; Mathew, N.R.; Mathews, B.; Mathias, J.H.; Mathieson, E.; Matthew, M.; Matthews, F.; Matthews, L.; Matthews, M.G.; Mawson, M.; Maybury, G.W.; Mayhow, S.J.; McAllister, C.; McCallum, A.; McCalum, J.; McCleary, J.; McCrae, A.W.R.; McDonald, D.; McFadzean, S.; McGibbon, R.; McHugh, R.; McKinnell, J.; McLaughin, J.; McLean, I.; McWilliam, S.J.; Meadows, N.P.; Mearns, C.; Meek, P.; Mellor, D.; Mellor, J.R.;

Melville-Stephens, J.; Mendel, H.; Menzies, I.S.; Merritt, W.; Merseyside Natural History Survey; Middle Thames Natural History Society; Middleton, W.; Miles, P.M.; Millar, R.D.; Miller, H.P.W.; Miller, J.; Miller, K.W.; Millett, M.; Mills, P.; Millward, A.; Milne-Redhead, M.; Milton, P.; Mitchell, S.; Mobsby, P.; Moffet, A.; Molville, S.; Montgomery Canal Survey; Moon, A.; Moore, I.; Moors River Survey Team; Moran, S.A.; Moreby, C.; Moreton, B.D.; Morgan, I.K.; Morgan, M.J.; Morgan, R.W.; Morison, G.D.; Morley, C.; Morris, B.C.; Morris, M.G.; Moseley, M.E.; Moss, E.; Mount, G.M.C.; Muggleton, J.; Murchison, H.; Murdoch, D.; Murphy, D.; Murphy, J.E.; Murphy, M.; Murphy, M.D.; Murphy, R.P.; Murphy, T.E.; Murse, E.; Musson, D.; Nathan, L.; Nature Conservancy Council + NCC East Anglian Fens Invertebrate Survey + NCC Invertebrate Site Register; Nau, B.S.; Naylor, G.; Neal, C.; Neil, C.J.; Nelson, B.; Newbold, C.; Newcombe, M.J.; Newell, S.; Newell, S.C.; Newman, B.H.; Newman, P.M.; Newton, A.; Newton, J.M.; Newton, R.J.; Nicholls, M.J.; Nicholls, S.; Nobes, G.; Noble, F.A.; Norfolk Young Naturalists; Norgate, F.; North Kent Wildlife Preservation Society; North, M.; O'Conner, J.P.; O'Connor, M.A.; O'Flanagan, C.D.; O'Mahony, E.; O'Mahony, J.; O'Neill, M.A.; O'Rinden, W.P.; O'Toole, C.; Oates, M.; Oates, R.; Ogilvie, M.A.; Okely, E.; Oldham Environmental Health Department; Oldham Industrial Co-op Society; Oldham Police; Oldroyd, D.; Ollerearnshaw, J.; Onslow, N.; Orchart, E.; Osborne, P.J.; Ovenden, D.W.; Owen, C.E.; Oxford University Museum, Hope Department; Oxford, G.; Packer, L.; Page, M.; Pallant, D.; Palmer, C.; Palmer, C.J.; Pannell, F.A.; Park, M.A.; Parker, A.; Parker, B.N.; Parker, J.; Parker, R.; Parker, R.H.; Parker, W.E.; Parmenter, L.; Parr, A.; Parry, D.; Parsons, A.J.; Parsons, M.; Parsons, R.M.; Partridge, D.; Passmore Edwards Museum; Patmore, J.M.; Patol, C.; Paton, J.A.; Paton, V.S.; Paul, C.; Paul, J.; Pavett, P.M.; Payne, K.; Payne, R.; Payne, R.G.; Payne, R.M.; Payne, R.W.; Peachey, C.; Pearce, E.J.; Pearce, J.; Pedley, I.; Peel, H.A.C.; Peet, T.N.D.; Pelham-Clinton, E.C.; Penhallurick, R.D.; Pennington, E.; Pepin, C.; Percival, N.; Perkins, J.F.; Perkins, R.C.L.; Perring, F.H.; Petrie, A.B.; Phillips, J.; Phillips, R.A.; Phillips, R.M.; Phillips, W.M.; Philp, B.; Philp, E.G.; Pickard, A.C.; Pickard, B.C.; Pickard, C.M.; Pickess, B.; Pickles, M.E.; Pilkington, G.; Pinchen, B.J.; Pitkin, B.R.; Pitkin, L.M.; Pitt, R.; Plant, C.W.; Ponting, A.F.; Pool, C.J.; Poole, A.; Poole, C.; Poole, J.; Porritt, G.T.; Porter, J.; Porter, K.; Preddy, S.; Price, B.E.H.; Price, E.B.; Price, S.; Price, S.C.; Price, T.; Prince, A.J.; Prince, T.; Prior, G.; Procter, B.A.; Pryce, R.D.; Puckett, J.; Pullen, G.; Pyman, G.A.; Pyner, A.K.; Quest, R.; Rae, B.; Ragge, D.R.; Ragge, N.; Ragge, R.G.M.; Ramsay, A.; Ramsden, N.; Randolph, S.; Rands, D.G.; Rands, E.B.; Ranger, J.; Read, M.; Reading Museum; Reavey, D.;

Redgate, N.; Redman, B.; Redshaw, E.J.; Reeve, K.M.; Reeve, P.J.; Reid, D.; Reid, J.H.; Reid, S.; Rey, S.; Reynolds, F.P.J.; Reynolds, K.; Reynolds, W.J.; Rheinallt, T.Ap; Rhodes, F.; Rhodes, J.D.; Richards, I.J.; Richards, J.P.; Richards, O.W.; Richards, T.J.; Richardson, A.; Richardson, D.T.; Richardson, J.A.; Richmond, A.; Richmond, D.I.; Richmond, J.L.; Richmond, R.; Richmond, R.M.; Richmond, S.E.; Ridgway, A.; Ritchie, J.M.; Ritchie, M.; Rivers, C.F.; Rix, S.; Robbins, J.; Roberts, B.; Roberts, F.; Roberts, J.E.H.; Roberts, M.; Roberts, M.J.; Roberts, S.; Roberts, W.; Robertson, D.A.; Robertson, J.S.; Robertson, L.R.; Robertson, R.B.; Robinson, G.F.B.; Robinson, R.; Robson, L.; Rodger, A.M.; Roe, D.; Roe, H.E.; Roebuck, W.D.; Rogers, R.; Roper, C.M.T.; Rose, A.; Rose, A.J.; Rose, E.; Rose, F.; Rotheray, G.; Rouse, S.; Rowden, A.O.; Rowe, B.L.; Rowe, V.A.; Rowland, K.; Rowley, B.; Ruck, O.L.; Rudkin, P.; Ruffell, R.D.; Rundle, A.J.; Russell, R.D.; Rutter, G.; Ruttledge, R.F.; Rylands, K.; Ryle, G.B.; Sampson, D.; Samuel, A.; Samuel, D.; Samways, M.J.; Sanders, W.J.; Sankey, J.H.P.; Sargent, H.B.; Saunders, D.; Saunders, D.R.; Saunders, I.; Saunders, K.; Saville, B.B.; Scott, D.W.; Scott, H.C.; Scrivens, S.; Searle, C.A.; Seaward, A.; Service, M.; Seymour, P.; Shaffer, M.; Shardlow, M.E.A.; Sharp, D.; Sharples, R.; Shaughnessy, J.P; Shaw, E.; Shaw, S.; Sheppard, C.C.; Sheppard, D.A.; Sherwood, R.; Shetton, S.W.; Shillito, J.F.; Shirley, P.R.; Shirt, D.B.; Shorock, H.; Shotton, F.W.; Shreeves, G.; Side, K.C.; Silvester, E.; Simms, C.; Simpson, A.N.B.; Sims, G.; Sims, R.B.; Skelton, M.J.L.; Skidmore, P.; Skinner, A.; Skinner, D.R.; Skinner, G.J.; Slator, C.; Sloane, R.H.; Smart, J.; Smith, A.V.; Smith, D.A.; Smith, D.H.; Smith, D.J.; Smith, G.; Smith, G.A.; Smith, G.S.; Smith, I.; Smith, I.F.; Smith, K.G.V.; Smith, P.H.; Smith, R.E.; Smith, W.E.; Smout, P.; Smout, R.; Snell, C.A.; Solman, D.; Sorby Natural History Society; South, R.; Southwood, T.R.E.; Spalding, A.; Speight, M.C.D.; Spooner, B.; Spooner, M.; Spouge, V.; St Ivo School Entomological Society; Stamp, N.; Stanfield, J.H.; Stanley, M.; Steeden, C.D.; Steeden, C.F.; Steeden, N.J.; Steel, W.O.; Steer, J.B.; Steer, John.; Stelfox, A.W.; Sterry, D.; Stevens, R.A.; Stewart, N.F.; Stock, J.A.; Stock, P.M.; Stokes, F.G.; Stone, D.A.; Stone, S.; Stowland, I.; Streeter, D.T.; Stroud, H.; Stroud, J.A.; Stubbs, A.E.; Sturmer, K.R.C.; Stuttard, P.; Summers, Mrs; Summerson, F.C.; Sumner, Miss.; Surry, R.J.; Sussex, D.; Sutherland, W.J.; Sutton, P.; Sutton, S.L.; Swain, A.M.; Swire, P.W.; Swynnerton, B.F.A.; Swynnerton, C.F.M.; Tabor, R.C.; Tagg, D.; Tailby, T.W.; Tameside Health Department; Tamms, W.A.; Tannett, P.G.; Taplin, J.; Tate and Lyle; Taylor, E.; Taylor, M.A.; Taylor, R.; Taylor, W.T.; Teagle, J.F.; Teagle, W.G.; Telfer, D.; Telfer, M.G.; Telfer, S.; Temple, V.; Tew, I.; The Natural History Museum, London; Thomas, A.C.;

Thomas, D.L.; Thomas, G.; Thomas, G.P.; Thomas, J.; Thomas, J.A.; Thomas, Mrs; Thomas, P.; Thomas, R.J.; Thompson, R.G.; Thompson, W.; Thornley, A.; Thornton, M.; Thorpe, M.; Thosen, G.; Tillotson, I.J.L.; Tilmouth, J.E.; Timmins, C.J.; Tinning, P.C.; Tittensor, A.M.; Tolhurst, D.J.; Tomlinson, R.; Toms, P.; Tottenham, C.E.; Townsend, B.C.; Townsend, C.C.; Tozer, D.; Trayner, L.; Trayner, M.; Trebble, N.; Tremewan, W.G.; Treseder, N.G.; Trought, T.E.; Tubbs, C.R.; Tuck, M.; Tucker, K.T.; Turk, F.A.; Turk, S.M.; Turnbull, N.; Turner, G.; Turner, H.J.; Turner, P.F.; Turpin, R.; Turpin, S.J.; Tyler, M.W.; Uffen, R.W.J.; Underwood, F.R.; Vane-Wright, R.I.; Vardy, C.R.; Varley, G.C.; Vaughan, A.; Vaughan, I.M.; Vaughan, O.; Vaughan-Jones, A.; Veall, M.E.; Veevers, D.J.; Verdcourt, B.; Vesey-Fitzgerald, B.; Vorkerne, G.J.; Vrendenburg, G.V.; Wagner, C.; Wagstaffe, R.; Wake, A.J.; Wake, A.K.; Wakefield, B.; Walker, I.; Walker, J.J.; Wall, C.; Wall, G.; Wallace, B.; Wallace, E.C.; Wallace, I.D.; Wallace, J.; Wallace, T.J.; Walsh, G.B.; Walters, J.; Walters, M.P.; Walton, J.S.; Ward, J.; Ward, P.H.; Wardens, V.C.P.; Waring, P.; Warne, A.C.; Warwick Wildlife Trust; Warwick, S.; Wash, R.J.; Watchman, A.; Waterhouse, C.O.; Waterston, A.R.; Waterston, J.; Watkins, M.A.; Watkins, O.G.; Watson, A.R.; Watson, S.M.; Watt, K.R.; Weaver, M.M.; Webb, N.R.; Webb, P.C.; Webb, P.G.; Webster, D.; Welch, R.C.; Welstead, A.R.; Welstead, N.I.; West, W.; Westcott, R.J.K.; Whale, K.; Wheatley, H.M.; White, D.; White, F.B.; White, H.V.; White, J.; White, P.; White, R.L.; Whitehead, J.E.; Whitehead, P.F.; Whitehead, R.W.; Whiteley, D.; Whybrow, E.; Whyn, B.; Widgery, C.R.; Widgery, J.; Widgery, K.J.; Widgery, P.E.; Wildig, A.W.; Wilkinson, T.; Wilkinson, W.; Williams, B.; Williams, D.; Williams, G.E.; Williams, I.; Williams, L.; Williams, L.H.J.; Williams, M.de Courcy; Williams, R.; Williamson, K.; Willis, J.; Willott, J.; Wilson, J.; Wilson, T.J.; Wiltshire Biological Records Centre; Wincott, S.C.; Wise, A.J.; Wisniewski, P.J.; Wood, A.D.; Wood, A.H.; Wood, J.; Wood, N.D.; Woodhams, M.G.; Woodroffe, G.E.; Woodruff, E.R.; Woods, R.G.; Woodward, J.; Worthington, E.B.; Wright, A.; Wright, R.V.; Wright, S.; Wyatt, L.B.; Wychwood Survey; Wyhed, P.R.; Wytham Survey; Yalden, D.W.; Yates, B.; Yerbury, J.W.; Young, R.; Zealand, K.B.; Zeuner, F.E.; Zitman, D..

References

Ball, S.G. 1986. *Terrestrial and freshwater invertebrates with Red Data Book, Notable or Habitat Indicator status.* (Invertebrate Site Register Report no. 66.) Peterborough: Nature Conservancy Council.

Ball, S.G. 1994. The Invertebrate Site Register - objectives and achievements. In: *Invertebrates in the landscape: invertebrate recording in site evaluation and monitoring*, edited by P.T.Harding. *British Journal of Entomology and Natural History*, **7**, Supplement 1, 2-14.

Bellmann, H. 1988. *A field guide to the grasshoppers and crickets of Britain and northern Europe.* London: Collins.

Binet, L. 1954. *Le roman de la mante religieuse.* Rennes: Oberthur.

Brock, P.D. 1987. A third New Zealand stick insect (Phasmatodea) established in the British Isles, with notes on the other species, including a correction. In: *Proceedings of the 1st International Symposium on stick insects*, edited by M. Mazzini & V. Scali, 125-132. Siena: Centrooffset Siena.

Brock, P.D. 1991. *Stick insects of Britain, Europe and the Mediterranean.* London: Fitzgerald Publishing.

Brown, V.K. 1983. *Grasshoppers.* (Naturalist's Handbooks 2.) Cambridge: Cambridge University Press. [Revised 1990 with new publisher - Slough: Richmond Publishing.]

Brown, V.K. 1978. Variations in voltinism and diapause intensity in *Nemobius sylvestris* (Bosc) (Orthoptera: Gryllidae). *Journal of Natural History*, **12**, 461-472.

Burr, M.D. 1897. *British Orthoptera.* Huddersfield: Huddersfield Economic and Educational Museum.

Burr, M.D. 1899. The Orthoptera of the Channel Islands. *Entomologists Record and Journal of Variation*, **11**, 245-246.

Burr, M.D. 1936. *British grasshoppers and their allies, a stimulus to their study.* London: Janson. [Originally printed in a limited edition by Philip Allan & Co.]

Burton, J.F. 1971. Some notes on Welsh Orthoptera with some first county records. *Nature in Wales*, **12**, 267-268.

Burton, J.F. 1990. The mystery of the Isle of Man's endangered grasshopper. *British Wildlife*, **2**, 37-41.

Cherrill, A.J. 1994. The current status of the lesser mottled grasshopper, *Stenobothrus stigmaticus* (Rambur) on the Isle of Man. *British Journal of Entomology and Natural History*, **7**, 53-58.

Cherrill, A.J. & Brown, V.K. 1990a. The life cycle and distribution of *Decticus verrucivorus* (L.) within a chalk grassland in southern England. *Biological Conservation*, **53**, 125-143.

Cherrill, A.J. & Brown, V.K. 1990b. The habitat requirements of adults of the wart-biter *Decticus verrucivorus* (L.) (Orthoptera: Tettigoniidae) in southern England. *Biological Conservation*, **53**, 145-157.

Cherrill, A.J. & Brown, V.K. 1991a. Colour variation in *Decticus verrucivorus* (L.) in southern England. *Entomologist's Gazette*, **42**, 175-183.

Cherrill, A.J. & Brown, V.K. 1991b. The effects of the summer of 1989 on the phenology of the wart-biter *Decticus verrucivorus* (L.) (Orthoptera: Tettigoniidae) in Britain. *British Journal of Entomology and Natural History*, **4**, 163-168.

Cherrill, A.J. & Brown, V.K. 1992. Ontogenetic changes in microhabitat preferences of *Decticus verrucivorus* (L.) (Orthoptera: Tettigoniidae) at the edge of its range. *Ecography*, **15**, 37-44.

Cherrill, A.J., Shaughnessy, J. & Brown, V.K. 1991. Oviposition behaviour of the bush-cricket *Decticus verrucivorus* (L.) (Orthoptera: Tettigoniidae). *Entomologist*, **110**, 37-42.

Copson, P. 1984. *Distribution atlas: orthopteroids in Warwickshire.* Warwick: Warwick Museum.

Cotton, D.C.F. 1980. Distribution records of Orthoptera (Insecta) from Ireland. *Bulletin of the Irish Biogeographical Society*, **4**, 13-22.

Cotton, D.C.F. 1982. A synopsis of the Irish Orthoptera. *Entomologist's Gazette*, **33**, 243-254.

Cropper, R.S. 1993. A survey of the distribution of Orthoptera and allied insects in south and north Somerset. *Proceedings of the Somerset Archaeological and Natural History Society for 1991*, **135**, 213-226.

Davies, M. 1987. Grasshoppers, crickets and bush-crickets in Devon. *Nature in Devon*, **8**, 45-64.

Department of Environment. 1994. *Biodiversity: the UK action plan.* (CM 2428.) London: HMSO.

Devriese, H. 1988. *Saltatoria Belgica. Voorlopige verspreidingsatlas van de sprinkhanen en krekels van België.* Brussels: Institut Royal des Sciences naturelles de Belgique.

Evans, W. 1901. A contribution towards the list of Scottish Orthoptera. *Annals of Scottish Natural History*, **37**, 26-31.

Forsyth, J. 1968. Request for records of grasshoppers, crickets and cockroaches. *Irish Naturalists' Journal*, **16**, 81.

Foss, P.J. & Speight, M.C.D. 1989. *Stethophyma*

grossum (L.): a further midland record with comment on the status of this grasshopper in Ireland (Orthoptera: Acrididae). *Bulletin of the Irish Biogeographical Society*, **12**, 90-93.

Fowles, A.P. 1986. Crickets and grasshoppers in Dyfed. *Dyfed Invertebrate Group Newsletter*, **1**, 7-12. [Subsequent updates have been published in the same journal.]

Fowles, A.P. 1992. *Provisional distribution maps for the Orthoptera of Ceridigion.* Bangor: Countryside Council for Wales.

Frost, R.A. 1991. The status of grasshoppers and crickets in Derbyshire. *Journal of the Derbyshire Entomological Society*, Autumn, 12-21. [Subsequent updates have been published in the same journal.]

Gillham, M.E. 1977. *The natural history of Gower.* Cowbridge: D. Brown.

Good, J.A. & Cullinane, D. 1990. The great green bush-cricket, *Tettigonia viridissima* L. (Orthoptera: Tettigoniidae) imported in a tent. *Irish Naturalists' Journal*, **23**, 220.

Haes, E.C.M. 1976. Orthoptera in Sussex. *Entomologist's Gazette*, **27**, 181-202.

Haes, E.C.M., ed. 1979. *Provisional atlas of the insects of the British Isles. Part 6. Orthoptera.* (2nd edn.) Huntingdon: Biological Records Centre.

Haes, E.C.M. 1990. *Grasshoppers and related insects in Cornwall.* (Cornish Biological Records, 12.) Redruth: Cornish Biological Records Unit.

Haes, E.C.M., Cherrill, A.J. & Brown, V.K. 1990. Meteorological correlates of the abundance of *Decticus verrucivorus* (L.) (Tettigoniidae) in Britain. *Entomologist*, **109**, 93-99.

Harding, P.T. & Sheail, J. 1992. The Biological Records Centre: a pioneer in data gathering and retrieval. In: *Biological recording of changes in British wildlife*, edited by P.T.Harding, 5-19. (ITE symposium no.26.) London: HMSO.

Holst, K.T. 1986. *The Saltatoria (bush-crickets, crickets and grasshoppers) of northern Europe.* (Fauna Entomologica Scandinavica, vol. 16.) Klampenborg: Scandinavian Science Press.

Ingrisch, S. 1979. *Fundortkataster der Bundesrepublik Deutschland - Teil 13: Regionalkataster des Landes Hessen - Die Orthopteren, Dermapteren und Blattopteren (Insects: Orthoptera, Dermaptera, Blattoptera) von Hessen.* Saarbrücken und Heidelberg: Universität des Saarlandes.

Kevan, D.K.McE. 1952. A summary of the recorded distribution of British orthopteroids. *Transactions of the Society for British Entomology*, **11**, 165-180.

Kevan, D. K. McE. 1956. The known distribution of British orthopteroids, fourth supplement. *Journal of the Society for British Entomology*, **5**, 187-192.

Kevan, D.K.McE. 1961. A revised summary of the known distribution of British orthopteroids. *Transactions of the Society for British Entomology*, **14**, 187-205.

Kleukers, R.M.J.C., van Nieukerken, E.J., Odé, B., Willemse, L.P.M. & van Wingerden W.K.R.E. 1997. *De sprinkhanen en krekels van Nederland (Orthoptera).* Nederlandse Fauna 1. Leiden: Nationaal Natuurhistorisch Museum, Uitgeverij KNNV & EIS-Nederland.

Lee, M. 1995. A survey into the distribution of the stick insects of Britain. *Phasmid Studies*, **4**, 15-23.

Le Sueur, F. 1976. *A natural history of Jersey.* Chichester: Phillimore.

Lucas, W.J. 1920. *A monograph of the British Orthoptera.* London: Ray Society.

Luff, W.A. 1896. The Orthoptera of Guernsey. *Report and Transactions of the Guernsey Society for Natural Science*, **3**, 113-117.

Mahon, A. 1992. The distribution of Orthoptera in Dorset. *Recording Dorset*, **2**, 2-9.

Marshall, J.A. 1974. The British Orthoptera since 1800. In: *The changing flora and fauna of Britain*, edited by D.L.Hawksworth, 307-322. (Systematics Association Special Volume no.6.) London: Academic Press.

Marshall, J.A. & Haes, E.C.M. 1988. *Grasshoppers and allied insects of Great Britain and Ireland.* Colchester: Harley.

McNamara, D. 1996. A note on *Bacillus rossius*, the Corsican stick insect. *Bulletin of the Amateur Entomologists' Society*, **55**, 31-32.

Merritt, R., Moore, N.W. & Eversham, B.C. 1996. *Atlas of the dragonflies of Britain and Ireland.* London: HMSO.

Paul, J. 1989. *Grasshoppers and crickets of Berkshire, Buckinghamshire & Oxfordshire.* Oxford: Pisces.

Paul, J. 1994. The Orthoptera of Jersey, Channel Islands. *Entomologist's Gazette*, **45**, 185-195.

Pickard, B.C. 1954. *Grasshoppers and crickets of Great Britain and the Channel Islands.* Ilkley. Privately published.

Pickard, B.C. 1956. The distribution of *Omocestus rufipes* (Zett.) in Great Britain, with descriptions of the two British species of the genus *Omocestus* (Bol., (Orth., Acrididae). *Entomologist's Monthly Magazine*, **95**, 51-53.

Ragge, D.R. 1963. First record of the grasshopper *Stenobothrus stigmaticus* (Rambur) (Acrididae) in the British Isles, with other new distribution records and notes on the origin of the British Orthoptera. *Entomologist*, **96**, 211-217.

Ragge, D.R. 1965. *Grasshoppers, crickets and cockroaches of the British Isles.* London: Warne.

Ragge, D.R. 1973. The British Orthoptera: a supplement. *Entomologist's Gazette*, **24**, 227-245.

Rands, D.G. 1978. The distribution of common bush-crickets and grasshoppers in Bedfordshire. *Bedfordshire Naturalist*, **32**, 25-30. [Subsequent updates have been published in the same journal.]

Richmond, D.I. 1995. *The grasshoppers and crickets of Norfolk: an update of records received to the end of 1994.* Reepham. Privately published.

Richmond, D.I. & Irwin, A.G. 1991. The grasshoppers and crickets of Norfolk. *Transactions of the Norfolk and Norwich Naturalists' Society*, **29**, 53-70.

Ryan, J.G., O'Connor, J.P. & Beirne, B.P. 1984. *A bibliography of Irish Entomology.* Glenageary: Flyleaf Press.

Salmon, J.T. 1992. *The stick insects of New Zealand.* Auckland: Reed.

Shaw, E. 1889. Synopsis of the British Orthoptera. *Entomologist's Monthly Magazine*, **25**, 354-359, 365-372, 409-421, 450-455.

Shaw, E. 1890. Synopsis of the British Orthoptera. *Entomologist's Monthly Magazine*, **26**, 56-64, 94-97, 167-176.

Shirt, D.B., ed. 1987. *British Red Data Books: 2. Insects.* Peterborough: Nature Conservancy Council.

Sibbald, R. 1684. *Scotia illustrata.* Edinburgh.

Skelton, M.J., ed. 1978. *Provisional atlas of the insects of the British Isles. Part 6. Orthoptera.* Huntingdon: Biological Records Centre.

Speight, M.C.D. 1976. Irish Orthoptera: some distribution records, including a first record of *Tachycines asynamorus* Adelung (Rhaphidophoridae). *Irish Naturalists' Journal*, **18**, 272-273.

Thorens, P. & Nadig, A. In prep. *Atlas de distribution des Orthoptères (Saltatoria) de Suisse.* Documenta faunistica helveticae. Neichâtel: Centre suisse de cartographie de la faune.

Timmins, C.J. 1994a. The population size of *Pseudomogoplistes squamiger* Fischer (Orth. Gryllidae) on Chesil Beach, Dorset. *Entomologist's Monthly Magazine*, **130**, 66.

Timmins, C.J. 1994b. The life cycle size of *Pseudomogoplistes squamiger* Fischer (Orthopt. Gryllidae). *Entomologist's Monthly Magazine*, **130**, 218.

Timmins, C.J. 1995. Parental behaviour and early development of Lesne's earwig *Forficula lesnei* (Finot) (Dermaptera: Forficulidae). *Entomologist*, **114**, 123-127.

Voisin, J.-F. 1992. *Atlas des Orthoptères de France. Etat d'avancement au 31-XII-1991.* Paris: Secrétariat de la Faune et de la Flore, Muséum National d'Histoire Naturelle.

Wake, A.J. 1984. *The grasshoppers and crickets of Essex. A provisional atlas.* Colchester: Colchester and Essex Museums.

Welstead, A.R. & Welstead, N.I. 1985. *Orthoptera of the New Forest and its environs.* Hythe. Privately published.

White, G. 1789. *The natural history and antiquities of Selborne.* London: B. White & Son.

Whiteley, D. 1981. Grasshoppers, crickets and cockroaches (Orthoptera) of the Sheffield area. *Sorby Record*, **19**, 68-75. [Subsequent updates have been published in the same journal.]

Widgery, J.P. 1991. A provisional commentary on the status of crickets, grasshoppers and related insects (Orthopteroids) in Hertfordshire. *Transactions of the Hertfordshire Natural History Society*, 18-24.

Zeuner, F.E. 1940. The Orthoptera Saltatoria of Jersey, Channel Islands. *Proceedings of the Royal Entomological Society of London* (B), **9**, 105-110.

Index to species accounts and distribution maps

Printed in the United Kingdom for The Stationery Office
N21899 C10 7/97